Recent Trends in BIOTECHNOLOGY

– Editor –

T. Pullaiah

Department of Botany

Sri Krishnadevaraya University

Anantapur - 515 003, Andhra Pradesh, India

and

N. Rama Swamy

Department of Botany

Kakatiya University

Warangal-506 009

Andhra Pradesh, India

2018

Regency Publications

A Division of

Astral International Pvt. Ltd.

New Delhi – 110 002

First Published, 2006
Reprinted, 2018

ISBN: 978-81-89233-22-8 (Harbound)

Publisher's note:
Every possible effort has been made to ensure that the information contained in this book is accurate at the time of going to press, and the publisher and author cannot accept responsibility for any errors or omissions, however caused. No responsibility for loss or damage occasioned to any person acting, or refraining from action, as a result of the material in this publication can be accepted by the editor, the publisher or the author. The Publisher is not associated with any product or vendor mentioned in the book. The contents of this work are intended to further general scientific research, understanding and discussion only. Readers should consult with a specialist where appropriate.
Every effort has been made to trace the owners of copyright material used in this book, if any. The author and the publisher will be grateful for any omission brought to their notice for acknowledgement in the future editions of the book.

Published by : **Regency Publications**
A Division of
Astral International Pvt. Ltd.
– ISO 9001:2015 Certified Company –
4736/23, Ansari Road, Darya Ganj
New Delhi - 110 002
Phone: 011-4354 9197, 2327 8134
E-mail: info@astralint.com
Website: www.astralint.com

Preface

We edited this book with the objective of analyzing the biotechnological approaches for improvement of medicinal and crop species. The book provides information on genome sequencing in microbes, androgenic haploids, and various aspects of plant tissue culture, secondary metobolites, biopesticides and mycotoxins.

The book "Recent trends in Biotechnology" contains contributions from various research institutes and universities of India. The book includes 14 research and review articles. The first article by K. Koteswara Rao is on Microbial genome sequencing. The next article is on androgenic haploidy in Solanum species by Venkataiah *et al.* and followed by *in vitro* propagation of *Syzygium alternifolium* by Rama Murthy and Savithramma. Shankarmurthy *et al.*, have dealt on Plant Regeneration from Root culture of *Embelia ribes*. Six articles on importance of growth regulators in plant tissue culture in various crop and medicinally important species written by Gayatri and Rama Gopal, Param Geetham *et al.*, Karthik Iyer *et al.*, Ramulu *et al.*, Ugandhar *et al.* and Aruna *et al.*, Sai Ram *et al.*, have written on Antihyperglycenic activity of leaf callus of *Gymnema sylvestre*. Giridhar and S.M. Reddy, Rama Krishna and Narayanaiah gave an account of mycotoxins problems in India and their management and Botanical pesticides respectively. Venkataiah *et al.*, have reviewed the Importance of Biotechnology in improvement of vegetable Solanaceous crops.

We are grateful to all the leading scientists who have contributed chapters clearly elucidating the recent developments in various fields.

We hope that this book containing the contributions from various scientists definitely will be useful to post graduate students, teacher and researchers engaged in various disciplines of Botany and Biotechnology.

Editors
T. Pullaiah
N. Rama Swamy

Contents

Microbial Genome Sequencing

*K. Koteswara Rao**

Microorganisms comprise about 60% of the Earth's biomass. The genetic, metabolic and physiological diversity of the microbial species is far greater than that found in plants and animals. But the diversity of the microbial world is largely unknown, with less than one half of 1% of the estimated 2–3 billion microbial species identified. Of those species that have been described, the biological diversity is extraordinary having adopted to grow under extremes of temperature, pH, salt concentration and oxygen levels (Fraser *et al.*, 2000). The identification of total number of functional genes in genome can be looked at the following angles depending upon gene expression.

1. Genome is composed of genes of an organism defined by the complete sequence of DNA. The number of genes can be identified directly by defining open reading frames.
2. Transcriptome: Gene expression results in the generation of multiple m-RNA's and the transcriptome is likely to be larger than the number of genes.
3. Proteome is the complete set of proteins corresponding to the m-RNA's in the transcriptome.

The institute of Genomic Research Maryland USA (TIGR) published the first complete genome sequence for a free-living microorganism, *Haemophilus influenzae* (2). Since then another 1570 microbial genome sequences have been published. *Escherichia coli, Bacillus subtilis,* thermophilic species, archaebacteria and the first eukaryote *Saccharomyces cerevisiae* are included in the first completed microbial genome projects. Till now 1380 viral, 19 Archaebacteria, 174 eubacteria, one algae, ten fungi and

* Head, Dept. of Biotechnology, College of Science, Gandhi Institute of Technology & Management, Visakhapatnam 530045.

three protozoa have been sequenced as per the data base information (3) (Table 1).

Table 1: Microbes with completed Genome Sequences

Type	*No of genomes sequenced*
Viral Genomes	1380
Archaea	19
Eubacteria	174
Eukaryotes:	
Algae	01
Fungi	10
Protozoa	03

Microbial Genome Size and DNA Content

Comparison of the results from the completed prokaryotic genome sequences, containing more than 50 Mb of DNA sequence and approximately 54,000 predicted open reading frames (ORF's), has revealed that gene density in the microbes is consistent in many species, with about one gene per kilobase. Almost half of the ORF's in each species are of unknown biological function (1). Genome size data from smallest microorganism Mycoplasma to Mammals Chimpanjee is shown in table 2.

Table 2: Variation in Genome Sizes from Bacteria to Mammal

Phylum	*Species*	*Genome (bp)*
Algae	*Pyrenomas salina*	6.6×10^5
Mycoplasma	*M. pneumoniae*	1.1×10^6
Bacteria	*Escherichia coli*	4.2×10^6
Yeast	*Sacchromyces cerevisiae*	1.3×10^7
Slime mold	*D. discoideum*	5.4×10^7
Nematode	*C. elegans*	8.0×10^7
Insect	*Drosophila melanogaster*	1.4×10^8
Bird	*G. domesticus*	1.2×10^9
Amphibian	*X. laevis*	3.1×10^9
Mammal	*Homo sapiens*	3.3×10^9

These results suggest algae and mycoplasms are the smallest prokaryotes. There is an increase in the minimum genome size from simple organisms to more complex prokaryotes and further to lower eukaryotes and later to much advanced eukaryotic system. Yeast genome size is not much higher size than that of big bacterial genomes. Further, increase in genome size higher than yeast to the tune of 5.4×10^7 bp is adequate to support the functioning of slime mold *D.discoideum*, a novel mold studied for cell differentiation. *C. elegans*, the first evolutionary series of multicellular nematode worm has a genome size of 8 × 10^7 bp. Further, it is necessary to have higher genome size in order to function their characteristic properties in case of insects, birds and mammals. Further studies on nature of DNA revealed that prokaryotes contains only non repetitive DNA. The lower eukaryotes have approximately 80% non repetitive DNA and the remaining 20% falls into repetitive DNA. The eukaryotic plant system differs from animal system where the former contains 80% of the genome is repetitive DNA where as the later has 50% only (4).

Bacterial Genomes

Prokaryotic genomes are very different from eukaryotic ones. There is some overlap between the largest prokaryotes and the smallest euakaryotic genomes. In general the trend is on the whole prokaryotic genomes are much smaller. It is interesting to note that prokaryotic genomes are even more compact than those of lower eukaryotes. it is obvious that there are more genes and less space between them. Some genes have virtually no space between them. The *E. coli* genome, for example is just 4.7 Mb, two-fifths of the yeast genome size. The sequence of the genomes of bacteria and archae show that virtually all of the DNA nearly 85–90% codes for RNA or protein. The genome size is proportional to the number of genes. Most of the bacteria with genome sizes below 1.5 to 1.8 Mb are obligate intracellular parasites with small molecules with minimum number of functions (3, 4 & 5). In case of obligate intracellular parasites, all types of functional genes are less in number compared to bacteria with larger genomes. Mycoplasma has the smallest genome with 470 functional genes. The archaea bacteria which has been placed in between prokaryotes and eukaryotes in classification, earlier it was thought that the genome size and gene number are almost similar to other sequenced bacteria. But now as per the latest microbial genome sequencing agencies it is not true (table 3). The recent data revealed their genomic size perfectly fitting their classification. For example *Methanococcus jannaschii, Methanopyrus kandleri, Aeropyrum premix, Methanobacterium thermoautotrophicum* have genome size almost equal to bacterial human pathogens like *Haemophilus influenzae, Helicobacter pylori,* which are of lower genome size, where as some other archae, *Methanosarcina acetivorans, Methanosarcina mazei* have higher range of genome size i.e. 5.7 and 4.0 Mb respectively showing similarity to most other bacteria and nearly half of the yeast genome size.

The wide spread human pathogen *Helicobacter pylori* is a major cause of gastric and duodenal ulcers and gastric cancer (5). The stomachs of more than half the people on earth are host to *H. pylori,* a corkscrew shaped microbe that may be mankind's most common bacterial infection. Studying parts of the genome has clarified how the microbe manages to thrive

Table 3: Sequenced Genomes from Representative Prokaryotic and Archae microorganisms (3, 6, 7)

Year	*Organism*	*Genome size (Mb)*	*No. of genes (Proteins)*
Prokaryotes			
1995	*Mycoplasma genitalium*	0.58	470
1995	*Aquifex acolicus*	1.5	1,512
1997	*Helicobacter pylori*	1.7	1,633
1995	*Haemophilus influenzae*	1.83	1,743
1998	*Deinococcus radiodurans*	2.65	2,580
2001	*Mycobacterium leprae*	3.3	(2,720)
2000	*Vibrio cholerae*	4.03	(3,827)
1995	*Bacillus subtilis*	4.2	4,100
1998	*Mycobacterium tuberculosis*	4.4	(3,918)
1997	*Escherichia coli*	4.7	4,288
Archae			
1996	*Methanococcus jannaschii*	1.66	1,738
2002	*Methanopyrus kandleri*	1.69	1,691
1997	*Methanobacterium thermoautotrophicum*	1.75	1,869
2002	*Methanosarcina mazei*	4.0	3,371
2002	*Methanosarcina acetivorans*	5.7	4,540

in acid in the stomach. Twenty-six of the 33 separated secreted proteins were identified by using proteome analysis (5). The identification of genes in prokaryotic genomes has advanced to the stage at which all protein coding regions can be identified with confidence. It is evident from the results presented in table 3 & 4 there is a good correlation between genome size and protein coding regions. Similar reports are available in earlier literature also (1).

Yeast Genome

In case of eukaryotic genomes, the relationship between genome size and gene number correlation do not exist. The genome of the typical yeasts *Saccharomyces cerevisiae* and *Saccharomyces pombe* were extensively studied (8). The size of *S.cerevisiae* nuclear genome is 11.6 Mb divided into 16 chromosomes which contain around 6310 genes and that of mitochondria around 0.085 Mb. It appears from the genome data, yeast genome is intermediate between prokaryote genome and higher eukaryote genome. *Dictyostelium discoidum*, the mold which was extensively exploited for fungal differentiation has three fold higher genome size i.e. 34.0 Mb with estimated 11,000 genes compared to yeast. In general, the simplest eukaryotes such as yeast and mold have relatively smallest genomes, while the higher eukaryotes such as vertebrates and higher plants have the largest genomes. *S. cerevisiae*

Table 4: Sequenced Genomes from Bacteriophages, Plasmids and Eukaryotic Nuclear and Organellells

Year	*Organism*	*Group*	*Genome size (Mb)*	*No of genes (Proteins)*
Eukaryotic nucleus				
1997	*Sacchromyces cerevisiae*	Yeast	11.6	6,310
2002	*Sacchromyces pombe*	Yeast	12.6	4,929
	Dictyostelium discoidum	Mold	34.0	11,000 est
Eukaryotic organelles				
1999	*Sacchromyces cerevisiae* Mitochondria	Yeast	0.085	43
Plasmids				
2000	F Plasmid	In *E. coli*	0.10 (C)	117
	Kalilo	In *Neurospora*	0.009 (L)	2
Bacterial virus				
1975	$\phi \times 174$	Bacterial virus	0.005	13
1999	T4	Bacterial virus	0.17 (L/C)	288

C = circular DNA; L – Linear DNA; L/C = Linear in free virus, Circular in cell

detailed genome analysis revealed that 5855 genes for proteins, 140 genes for rRNA, 275 genes for tRNA and 40 genes for snRNA (3, 6, 7).

Plasmids

Some information is available on plasmids. F plasmid in *E.coli* has the genome size of 0.10 Mb with 117 genes. *Neurospora crassa,* Kalilo plasmid has 0.009Mb genome size and 2 genes.

Bacterial Viruses

Until now total 1380 viruses have been sequenced and used for the classification.

Refrences

1. Claire M. Fraser, Jonathan A. Eisen and Steven L. Salzberg. 2000. *Nature* 108, 799–803.
2. R.D. Fleischmann. 1995. Science **269,** 496–512.
3. *www.ncbi.nlm.nih.gov* National Centre for Biotechnology Information (NCBI)
4. B. Lewin. 2004. Genes VIII. Pearson Prentice Hall.
5. T.A. Brown. 1999. Genomes: John Wiley & Son.
6. *www.genomesonline.org*
7. *www.genomenewsnetwork.org*
8. Saccharomyces genome database (SGD).
9. D. Bumann, S. Aksu, M. Wendland, K. Janek, U.Z. Arndt, N. Sabarth, T.E. Meyer and P.R. Jungblut. 2002. Infection and Immunity **70,** 3396.

Androgenic Haploidy in *Solanum* Species
A Review

P. Venkataiah, M. Praveen, M. Upender, M. Rambabu, T. Ugandhar, K. Subhash and *N. Rama Swamy**

ABSTRACT

The genus *Solanum* L. is economically important as several species are source of food, fodder, vegetables and drugs. Some species are grown in gardens as ornamentals. *In vitro* androgenesis plays a significant role for developing haploids in large numbers. This technique is also useful in reducing the time taken for production of homozygous double haploid line. Androgenic haploids are produced using anther and/or pollen culture *in vitro* in *Solanum* species. These haploids have been developed through embryogenesis/callus/direct plantlet formation in *Solanums* viz., *S. acaule, S. dulcamara, S. carolinense, S. chacoense, S. mamosum, S. melongena, S. nigrum, S. phureja, S. surattense, S. tuberosum* and *S. verrucosum* etc. Haploids derived from androgenesis have considerable potential for breeding experiments with agronomically important traits in these species. The present review summarizes the androgenic haploidy in *Solanum* spp. and its importance in breeding.

Introduction

Androgenic haploids are produced using anther and pollen culture *in vitro* in plants. Haploid plants develop directly from pollen grains in culture, either through direct formation of embryos from pollen grains or formation of callus and subsequent plant regeneration. One of the very popular methods for production of haploids is through anther culture on artificial medium which leads to the growth of microspores into sporophytes. Double haploids (DH) are induced from these androgenic haploids by *in vitro* androgenesis which play role in plant breeding experiments. *In vitro*

* Plant Biotechnology Research Group, Department of Botany, Kakatiya University, Warangal 506 009. E-mail: swamynr_dr@yahoo.com

androgenesis was reported for the first time by Guha and Maheshwari (1964, 1966, 1967) in *Datura innoxia*. They reported the development of embryos from microspores in cultured anthers of *Datura*. Since then, considerable research has been directed towards improving the technique for economically important plants. This technique requires optimizing culture conditions which vary from genotype to genotype. Haploid plants derived from this technique can greatly reduce the time taken to obtain a homozygous line after their chromosomes doubling (Morrison and Evans, 1988).

Homozygous double haploid plants (DH) obtained through microspore androgenesis can be used to select breeding lines of pharmaceutically important plants for their improvement (Collins *et al.*, 1974; Depaepe *et al.*, 1977). Through androgenesis, it is very easy to produce haploid plants for the improvement of cultivars with respect to yield and resistance towards biotic and abiotic stress (Bajaj, 1983). Androgenic haploids have been developed mainly in cereals and vegetable crops (Foroughi-Wehr and Wenzel, 1989) and also in Solanaceous crops especially Solanums (Anand and Govindappa, 1979; Aziz *et al.*, 1999; Kashyap *et al.*, 2003; Rama Swamy *et al.*, 2006).

Importance of Solanums

Solanaceae is a large family containing 90 genera and over 2,200 species and distributed in tropical and temperate regions of the world. But the chief centres of their distribution are Australia as well as Central and South America. Many species are cultivated throughout India whereas others occurring chiefly in Himalayas and Southern and Eastern parts of India. Familiar plants of this family are potato (*Solanum tuberosum*), egg plant (*S. melongena*), Indian-night-shade (*S. nigrum*) and Kantakari (*S. surattense*), tomato (*Lycopersicon esculentum*), red pepper (*Capsicum frutescens*), tobacco (*Nicotiana tabacum*), *Datura stramonium*, etc. *Solanum* species are of great economic importance and include many plants that furnish foods and drugs.

The genus *Solanum* L. is economically very important as several species are source of food, fodder and drugs. The fruits of brinjal are consumed as a vegetable all over India. Leaves of Indian-night-shade (*S. nigrum*) have long been used in medicine for treatment of *Scrofulous* and *Dysuria* (Gopalaswamienger, 1951; Schery, 1952; Baily, 1949). Fruits are also used in medicine. The fruits of Kantakari (*S. surattense*) are used as vegetable and also in ayurvedic medicine. Many *Solanum* species have resistance to pathogens/pests: Fungal wilts (*Fusarium* wilt, *Verticillium* wilt and bacterial wilt); Fruit rot resistance (*Phomopsis vexans* and *Cercospora solani*); Fruit and shoot borer resistance (*L. orbonalis*); Root-knot nematode resistance (*Meleidogyne* spp. & Spider mite); other pests/pathogens (*Aphis gossypii, Tetranychuns urticae*, eggplant mosaic virus (Tymovirus) and Mycoplasma (see Kashyap *et al.*, 2003). Several species of *Solanum* produce glyco-alkaloids which on hydrolysis and removal of sugar residues yield steroidal alkaloids: solanine, solamargine and solasodine are widely distributed among the members of the genus. The aglycones of these glyco-alkaloides are solanidine ($C_{27}H_{43}NO$) and solasodine ($C_{27}H_{43}NO$). From the foregoing it is evident that many *Solanum* species are useful as vegetable and medicine. Hence, these species have been exploited *in vitro* thoroughly that the regeneration protocols for micro-propagation, somatic embryogenesis, haploid production, somatic hybridization and genetic transformation have been established (Ugandhar, 2002; Kashyap *et al.*, 2003).

Production of androgenic haploids using anther and/or pollen culture have been reported in a number of *Solanum* species (Table 1). Haploids were

Table 1: Induction of androgenic haploids in *Solanum* species using anther and pollen culture

S. No	*Species*	*Response*	*Reference*
1.	*Solanum acaule* ssp. *acaule*	Callus and Plants	Rokka *et al.*(1998)
2.	*S. aethiopicum*	Plants	Rizza *et al.* (2001, 2002)
3.	*S. dulcamàra*	Embryogenesis	Zenkteler (1973)
4.	*S. carolinense*	Callus and Plants	Reynolds (1990 a, b)
5.	*S. chacoense*	Callus and plants	Chani (1998), Boluarte-Medina and Veilleux (2002)
6.	*S. integrifolium*	Callus and plants	Rotino *et al.* (2001)
7.	*S. mamosum*	Callus and plants	Anand and Govindappa (1979)
8.	*S. melongena*	Callus, shoots and plants	Raina and Iyer (1973), Gu (1979), Dumas de Vaulx and Chambonnet (1982), Rotino (1996), Miyoshi (1996)
9.	*S. nigrum*	Callus and plants	Harn (1971, 1972)
10.	*S. phureja*	Callus and plants	Snider and Veilleux (1994), Chani (1998), Sirasak and Veilaeux (1998), Boluarte-Medina and Veilleux (2002)
11.	*S. phureja* X *S. chacoense*	Callus and plants	Chani (1998), Boluarte-Medina and Veilleux (2002)
12.	*S. surattense*	Callus and shoots, Embryogenesis	Prasad *et al.* (1998), Sinha *et al.* (1978, 1979)
		Callus, Shoot buds and Embryogenesis	Rama Swamy *et al.* (2006)
13.	*S. tuberosum*	Callus, Embryogenesis, plants	Sopory *et al.* (1978), Sopory and Tan (1979), Wengel *et al.* (1979), Simon and Peloqiun (1977), Uhrig and Salamini (1987), Singsit and Veilleux (1989), Sonniono *et al.* (1989), Rokka *et al.* (1996)
14.	*S. verrucosum*	Callus and Plants	Weatherhead and Henshaw (1979)

Induction of embryogenesis on anther culture medium + 1.0 mg/l NAA + 3.5 mg/l Kn.

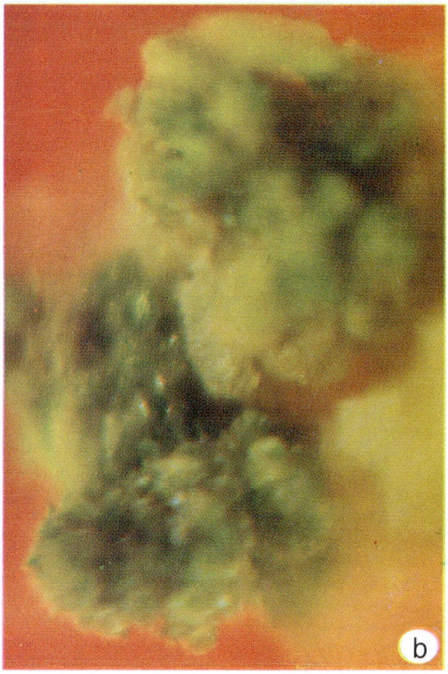

Androgenic calli grown on anther culture medium + 0.1 mg/l TDZ + 5.0 mg/l Kn (note the induction of multiple shoot buds)

Formation of multiple shoots after 30 days of culture on 0.1 mg/l TDZ + 4.0 mg/l Kn.

PLATE I

a–c: Anther culture in *Solanum surattense*

produced using the anther culture in wild Tomato (*S. carolinense*) by Reynolds (1990a). Reynolds (1990b) studied the interactions between calcium and auxin during pollen androgenesis in anther cultures of *S. carolinense*. Rokka *et al.* (1998) reported the production of androgenic dihaploid lines of the disomic tetraploid potato species *S. acaule* ssp. *acaule*. Zenkleter (1973) reported the *in vitro* development of embryos and seedlings of pollen grains of *S. dulcamara*.

Raina and Iyer (1978) were first to report plant regeneration from anther culture in eggplant. They regenerated homozygous diploid (double haploids) plants through callus developed from anthers that were previously treated with colchicine. Haploid plantlets were also produced using the same technique a year later in *S. melongena* (Isouard *et al.*, 1979).

Isouard *et al.* (1979) and Dumas de Vaulx and Chambonnet (1982) did an extensive work to improve the development of androgenic haploids in *S. melongena*. They observed that high temperatures (35 ± 2°C) incubation of anthers under dark conditions for the first 7–8 days improved the efficiency of haploid plant formation A temperature of 25°C produced 3.4 plants per 100 anthers while 8-day pretreatment at 35°C yielded an average of 12 plants per 100 anthers. They have also recorded the essentiality of both auxin and cytokinin during early stages of anther culture. The auxin in the media was found to be greatly influenced the ploidy level of the regenerated plants. With 2,4-D 15–50% regenerants were found to be diploid, while with NAA only haploid plants were regenerated. Similarly, Rotino *et al.* (1987) showed that haploid plant regeneration was affected by genotype, temperature, culture conditions, hormones and anther stage. A higher temperature governs the shift of the microspores from gametophytic stage to sporophytic stage. Rotino (1996) noted that the regeneration medium supplemented with same cytokinin showed the enhancement in the anther culture response. He also found the effect of stage of anther, genotype influence and culture conditions in androgenic haploid production of *S. melongena*. The effect of somaclonal variations on androgenic traits of embryogenic and androgenic colchicine-treated double haploid lines was also investigated (Rotino *et al.*, 1991). Plant height, fruit shape and yield were some of the features affected by somaclonal variations. Miyoshi (1996) cultured isolated microspores of *S. melongena* on Nitsch and Nitsch (1969) medium supplemented with NAA (0.5 mg/l) and BA (0.5 mg/l) initially for 4 weeks. The calli obtained were transferred to medium supplemented with Zeatin (4 mg/l) and IAA (0.2 mg/l) for plant regeneration. Culturing these microspores of egg plant in a sucrose-free medium was found to be effective to induce callus. Regeneration of shoots was also successful from these calli in a medium containing zeatin and IAA. The efficiency of haploid regeneration was additionally improved by preculturing anthers/or microspores for inducing plantlet regeneration (Gu, 1979; Miyoshi, 1996).

A limitation in potato breeding has been the unavailability of homozygous lines, bécause of self-incompatibility of most diploids (2n = 2x = 24) and in-breeding depression exhibited after selfing most genotypes, when possible, regardless of ploidy. Batty and Dunwell (1989) have found the efficacy of sucrose in anther culture of potato to stimulate embryo induction more than maltose. Sopory *et al.* (1978) have produced monoploid embryoids and plantlets in cultured anthers of *S. tuberosum*. The effect of sucrose, growth regulators and metabolic inhibitors on the development of pollen embryoids was studied in anther cultures of diploid potato (Sopory, 1977; Sopory and Tan 1979). Aziz *et al.* (1999a, b) have screened diploid *Solanum* genotypes responsive to different anther culture conditions and ploidy assessment of anther-derived roots and plantlets. They have also developed Randomly amplified polymorphic DNA (RAPD) markers using the amplification of DNA from single pollen grains of diploid (2n = 2x = 24) potato.

Two diploid (2n = 2x = 24) backcross potato populations (PBCp, and CBC) (CBC — backcross between CP2 and *S. chacoense* 80-1; PBCp-backcross between *S. phureja* 1–3 and CP2) were characterized for anther culture response (ACR). PBCp (*S. phureja* genotype 1–3 × CP2) and CBC (CP2 × *S. chacoense* Bitt. genotype 80-1) resulted from a cross between CP2 (intermediate ACR) and its parents, *S. chacoense* 80-1 (low ACR) and *S. phureja* 1–3 (high ACR). Three components of ACR were initially investigated: embryos per anther (EPA), embryo regeneration rate and percent monoploids (2n = 1x = 12) among regenerants. EPA was selected for further characterization because of its relative stability (Aziz *et al.*, 1999a, b). Plantlets were also obtained from anther and pollen calli of dihaploid potato (Sopory and Tan, 1979). Sirasak and Veilleux (1998) studied the effects of colchicine on androgenesis and ploidy of anther-derived plants and were examined in three experiments in diploid potato (*S. phureja*). They have recorded 62% monoploids among the regenerated plants.

Anther culture experiments were conducted to construct a mon oploid family. The donor plants used were hybrids between high leptine producing selections of *S. chacoense* and anther culture responsive selections of *S. phureja*. Several steps of the anther culture process were studied. The results indicated that genotype remains the main factor affecting anther culture response. Growing anther-donor plants in higher greenhouse temperatures (30°C day/20°C night) increased the number of EPA by 40 percent. A heat shock given to anthers in culture for 12 h at 35°C was also found to be beneficial resulting in an increase of the anther culture response by 40 percent. However the effect of the high temperature shock resulted in lower regeneration rates (Chani, 1998).

Harn (1971, 1972) developed the haploid plants through anther culture in *S. nigrum*. Callus induction and plantlet regeneration was observed in *in vitro* culture of excised anthers of *S. mamosum* (Anand and Govindappa, 1979). Weatherhead and Henshaw (1979) reported the production of homozygous diploid plants in *S. verrucosum*.

Prasad *et al.* (1998) have reported the androgenic stable somaclonal variant (tetraploid) in *S. surattense*. Rama Swamy *et al.* (2006) have obtained the androgenic haploids in medicinally important herb *S. surattense* (Plate Ia–c). They found the requirement of both auxins and cytokinins when the anthers cultured on modified Nitsch medium as it was observed by Prasad *et al.* (1998). The same phenomenon was also observed in *S. melongena* (Dumas de Vaulx and Chambonnet, 1982). The embryogenic callus induction was observed on 1.0 mg/l NAA + 3.0 mg/l and 3.5 mg/l Kn in *S. surattense* (Plate Ia). Multiple shoot bud proliferation was observed from embryogenic calli on 3.0–5.0 mg/l Kn (Plate Ib, c). Kn had showed the superiority in inducing more number of shoots compared to BAP in all the concentrations tested in combination with TDZ (0.1 mg/l) in anther cultures of *S. surattense* (Rama Swamy *et al.*, 2006). Similarly, Prasad *et al.* (1998) have also found the differentiation of progressively more shoot buds with its increasing concentrations from 0.25 to 3.0 mg/l BAP in anther cultures of *S. surattense*.

However, the success in anther culture technology seems to be highly dependent on pretreatment period, genotype, environment (temperature, light intensity, photoperiod), culture conditions, physiology/growth of anther-donor plant and their interactions in addition to the different types of media (Rotino, 1996).

Application of Androgenic Haploidy in Plant Breeding

Application of androgenic haploids in plant breeding programs is an established technique today. Haploids derived from anther cultures have considerable potential for plant breeding because of the time saved by the reduction of the classical selection cycle period and the genetic value of isogenic lines or homozygous diploids.

One of the major problems facing plant breeders is the simple fact that offspring from a cross have a great deal of variability for important traits. This problem can be solved through tissue culture to create haploid plants, whose chromosome numbers are then doubled to produce normal diploids, bearing two identical copies of every gene. Offspring of doubled haploid plants are therefore more genetically uniform, for all important traits. Homozygosity allows direct selection of recessive characters and simplified gametic ratios. The double haploid plants resulting from anther culture are useful for breeding in improving the efficiency with which superior genotypes can be identified and also the presence of a single set of chromosomes allows the detection of mutations controlled by recessive genes (Knapp, 1991; Mitchell *et al.*, 1992).

The double haploid plants have been successfully used in convetional breeding programs to obtain pure lines faster than selfed inbreds. Double haploid plants are homnozygous at all loci, and this may help to study the genetic basis of quantitative traits by overcoming the problems associated with the environmental variations. The double haploid plants have proven to be useful for breeding plants with useful agronomic traits, such as high yield, disease resistance, earliness, abiotic stress tolerance and other characters in various crops (Jacobsen and Sopory, 1978; Uhrig and Salamini, 1987; Waari, 1996).

Genetic maps of potato have been derived from compromised populations with unpredictable segregation ratios (Tanksley *et al.*, 1992; Jacobs *et al.*, 1995; Hosaka, 1999). Doubled-haploids developed by anther culture are used as inbred lines in mapping or in breeding programs (Wenzel *et al.*, 1979). However, for this to occur, potato clones must be competent for androgenesis. Dihaploid extraction by anther culture of superior tetraploid (2n = 4x = 48) clones has been limited (Rokka *et al.*, 1996); therefore, transferring this ability to commercially desirable clones or enhancing it would facilitate potato breeding at the diploid level.

Anther culture was successfully employed to reduce the ploidy of eggplant somatic hybrid between *S. melongena* and wild species (*S. integrifolium* and *S. aethiopicum* group Gilo) (Rotino *et al.*, 2001; Rizza *et al.*, 2001, 2002) to the diploid status and the corresponding plants may be more easily crossed with the diploid cultivated eggplant for transferring useful traits. It is generally accepted today that a step wise reduction in the ploidy level of potato offers the advantage of a simpler inheritance and also a better chance to combine qualitatively inherited characters (Wenzel *et al.*, 1979). Data now available demonstrate that field resistance to potato virus Y (PVY) and potato leaf roll virus (PLRV) was maintained and expressed at all ploidy levels over a 5 year period. When a genome has passed the monoploid level, a homozygous additive character can be transferred efficiently to the next generation via classical crossing. Similar results have been obtained for polygenic nematode resistance (Uhrig and Wenzel, 1987).

The recent approaches of Dihaploid (DH) lines are useful in genome mapping. In most cases, there is no immediate need to always locate the gene directly responsible for a character expressed at the phenotypic level. It would be sufficient for applied breeding strategies to detect characteristic restriction fragment length polymorphisms (RFLP) correlated to phenotypes. For such correlations, segregating plant populations are a necessary prerequisite. In a population of DH lines the identification of markers is much more secure, as most intermediate phenotypic expressions due to heterozygosity are excluded. A gene will segregate in a 1:1 ration for both the molecular marker and the phenotype at the plant level (Garner *et al.*, 1990).

Chani (1998) developed the SSR primer pairs in *S. phureja* which are useful in studying genetic relationships among cultivars and accessions in breeding programs. RAPD analysis was used in association with bulked segregant analysis to detect linkage with genes controlling leptine biosynthesis.

Though androgenic haploidy has potential value in plant breeding, but there is no much work on the genes involved in the anther culture response and as well as *in vitro* induction mechanism of anther/pollen culture. The anther culture response (ACR) is a two-phase process, development of microspores into embryos followed by conversion of embryos into plantlets. Simon and Peloquin (1977) also suggested the action of a dominant gene controlling callus regeneration from anthers. Uhrig and Salamini (1987) suggested the action of a dominant gene for androgenic competence. Singsit and Veilleux (1989) provided additional data to support the action of a single dominant gene controlling ACR in diploid potato hybrids. By contrast, Sonnino *et al.* (1989) suggested that several recessive genes controlled anther culture response as measured by EPA. Later, Meyer *et al.* (1993) proposed the involvement of no more than two genes for ACR. However, Singsit and Veilleux (1989) found no positive correlation between embryo formation and regeneration of embryos into plantlets, thus suggesting that these two characters were independently inherited. In contrast to potato, inbred lines are available for all these crops, making it less complicated to tag anther culture response. Cloning genes that control ACR may facilitate their integration by gene transfer techniques into desirable potato genotypes (Boluarte-Medina and Veilleux 2002).

Although, little is known about the gene action at the time of anther culture response, the mechanism of induction of *in vitro* androgenesis is still poorly understood. The androgenic induction requires some external stress to competent microspores. The stress may be in the form of heat, cold, chemicals, starvation or water stress. The stress response during induction involves some disruption and organization of the cytoskeleton, altering the polarity of the cytoplasmic components of competent microspores. Imposition of stress is correlated with the appearance of smH SP (small heat shock protein) transcripts in the affected cells. Without elicitation of the stress response, as indicated by the appearance of smH SP transcripts, androgenesis can't proceed (Babbar *et al.*, 2004). Thus, from the foregoing account it is clear that it needs further research to understand the mechanism of induction of *in vitro* androgenesis especially in Solanums.

References

Anand, V.V. and A.K. Govindappa. 1979. *In vitro* culture of excised anthers of *Solanum mamosum* L. *Indian J. Exp. Biol.*, **17**:444–445.

Aziz, A.N., J.E.A. Seabrook and G.C.C. Tai. 1999b. Amplification of RAPD markers from single pollen grains of diploid (2N = 2X = 24) potato. *Am. J. Potato Res.*, **76(4)**:179–182.

Aziz, A.N., J.E.A. Seabrook, G.C.C. Tai and H. De Jong. 1999a. Screening diploid *Solanum* genotypes responsive to different anther culture conditions and ploidy assessment of anther-derived roots and plantlets. *Am. J. Potato Res.*, **76(1)**:9–16.

Babbar, S.B., N. Kumari and J.K. Mishra. 2004. *In vitro* androgenesis: Events preceding its cytological manifestation. *In*: Plant Biotechnology and Molecular markers (ed.) P.S. Srivastava, Alka Narula and Sheela Srivastava, pp. 1–17, Anamaya Publishers, New Delhi.

Baily, L.H. 1949. Manual of cultivated plants. The Macmillan Co., New York.

Bajaj, Y.P.S. 1983. *In vitro* production of haploids. In: Handbook of Plant Cell Culture, Vol. 1. P.V. Ammirato, Y. Yamada, D.A. Evans, W.R. Sharp (eds.) New York: MacMillan, pp. 228–87.

Batty, N. and J. Dunwell. 1989. Effect of maltose on the response of potato in anther culture. *Plant Cell Tiss. Org. Cult.*, **18**:221–226.

Boluarte-Medina, T. and R.E. Veilleux. 2002. Phenotypic characterization and bulk segregant analysis of anther culture response in two backcross families of diploid potato. *Plant Cell Tiss. Org. Cult.*, **68**:277–286.

Chani, E. 1998. Molecular marker analysis of a segregating monoploid potato family, Ph.D. Thesis, Department of Horticulture, Virginia Polytechnic Institute & State University, Blacksburg, Virginia 24061, USA.

Collins, G.B., P.D. Legg and M.J. Kasperbauer. 1974. Use of anther-derived haploids in *Nicotiana*. I. Isolation of breeding lines differing in total alkaloid content. *Crop Sci.*, **14**:77–80.

Depaepe, R., C. Nitsch, M. Godard and J. Pernes. 1977. Potential from haploid and possible use in agriculture. *In*: Plant Tissue Culture and its Biotechnological Application, W. Barz, E. Reinhard, M.H. Zenk (eds.), Berlin: Springer-Verlag. pp. 341–52.

Dumas, de Vaulx, R. and D. Chambonnet. 1982. Culture *in vitro* d' anthers, dhaubergine (*Solanum melongena* L.) stimulation de la production de plantes an moyen, de traitments at + 36°C associeas a de faibles teneurs, substances de croissance. *Agronomie*, **2**:983–988.

Foronghi-Wehr, B. and G. Wenzel. 1989. Androgenic haploid production. IAPTC News Sletter, **58**:11–18.

Garner, A., H. Sieder, A. Jahoor, R.G. Herrmann and G. Wenzel. 1990. Assessment of the degree and the type of restriction fragment length polymorphism in barley (*Hordeum vulgare*). *Theor. Appl. Genet.*, **80**:826–832.

Gopalaswamienger, K.S. 1951. Complete gardening in India. The Hosali Press, Bangalore.

Gu, S.R. 1979. Plantlet formation from isolated pollen culture of eggplant (*Solanum melongena* L.). *Acta Bot. Sin.*, **21**:30–36.

Guha, S. and S.C. Maheshwari. 1964. *In vitro* production of embryos from anther of *Datura*. *Nature*, **204**:497.

Guha, S. and S.C. Maheshwari. 1966. Cell division and differentiation of embryos in the pollen grains of *Datura in vitro*. *Nature*, **212**:97–98.

Guha, S. and S.C. Maheshawari. 1967. Development of embryoids from pollen grains of *Datura in vitro*. *Phytomorphology*, **17**:454–461.

Harn, C. 1971. Studies on the anther culture in *Solanum nigrum*. *SABRAO Newsl.*, **3**:39–42.

Harn, C. 1972. Production of plants from anther of *Solanum nigrum* cultivated *in vitro*. *Caryologia*, **25**:429–437.

Hosaka, K. 1999. A genetic map of *Solanum phureja* genotype 1.22 constructed using RFLP and RAPD markers. *Am. J. Potato Res.*, **76**:97–102.

Isouard, G., C. Raquin and Y. Demarly. 1979. Obention de plantes haploids, et diploids par culture in vitro, anthers d aubergine (*Solanum melongena* L.). *C.R. Acad. Science Ser.* D., **288**:987–989

Jacobs, J.M.E., H.J. Van Eck, P. Arens, B. Verkerkbakker, B.T.L. Hekkert, H.J.M. Bastiaanssen, A. Elkharbotly, A. Pereira, E. Jacobsen and W.J. Stiekema. 1995. A genetic-map of potato (*Solanum tuberosum*) integrating molecular markers, including transposons, and classical markers. *Theor. Appl. Genet.*, **91**:289–300.

Jocobsen, E. and S.K. Sopory. 1978. The influence and possible recombination of genotypes on production of microspore embryoids in anther cultures of *Solanum tuberosum* and dihaploid hybrids. *Theor. Appl. Genet.*, **52**:119–123.

Kashyap, V., S. Vinod Kumar, C. Collonnier, F. Fusari, R. Haicour, G.L. Rotino, D. Sihachakr and M.V. Rajam. 2003. Biotechnology of eggplant. *Scientia Hort.*, **97**:1–25.

Knapp, S.J. 1991. Using molecular markers to map multiple quantitative trait loci models for back cross, recombinant inbred doubled haploid progeny. *Theor. Appl. Genet.*, **81**:333–338.

Meyer, R., F. Salamini and H. Uhrig. 1993. Isolation and characterization of potato diploid clones generating a high frequency of monohaploids or homozygous diploid androgenic plants. *Theor. Appl. Genet.*, **85**:905–912.

Mitchell, M.J., R.H. Busch and H.W. Rines. 1992. Comparison of lines derived by anther culture and single-seed descent in a spring wheat cross. *Crop. Sci.*, **32**:1446–1451.

Miyoshi, K. 1996. Callus induction and plantlet formation through culture of isolated microspores of eggplant (*Solanum melongena* L.). *Plant Cell Reports*, **15**:391–395.

Morrison, R.A. and D.A. Evans. 1988. Haploid plant from tissue culture: new plant variations in a shortened time frame. *Biotechnology*, **6**:684–690.

Nitsch, J.P. and C. Nitsch. 1969. Haploid plants from pollen grains. *Science*, **163**:85–87.

Prasad, R.N., M. Sharma, A.K. Sharma, H.C. Chaturvedi. 1998. Androgenic stable somaclonal variant of *Solanum surattense* Burm. f. *Indian J. Exptl. Biol.*, **36**:1097–1012.

Raina, S.K. and R.D. Iyer. 1973. Differentiation of diploid plants from pollen callus in anther cultures of *Solanum melongena* L.Z. Pflanzenzucht, **70**:275–280.

Rama Swamy, N., T. Ugandhar, M. Praveen, D. Sharada, M. Rambabu and M. Upender. 2006. Androgenic haploidy in Indian *Solanum*. *Plant Cell Tiss. Organ Cult.* (Communicated).

Reynolds, T.L. 1990a. Wild Tomato *Solanum carolinense*: Anther Culture and the Induction of Haploids. In: Biotechnology in Agriculture and Forestry, Vol. 12, Haploids in Crop Improvement-I, Y.P.S. Bajaj (ed.), Springer-Verlag, New York, Berlin, pp. 498–513.

Reynolds, T.L. 1990b. Interactions between calcium and auxin during pollen androgenesis in anther cultures of *Solanum carolinense*. *Plant Science*, **72**:109–114.

Rizza, F., F. Barchiensi, G. Mennela, M. Tacconi, C. Collonnier, D. Sihachakr, F. Fusari, N. Acciarri and G.L. Rotino. 2001. Production and characterization of dihaploid plants from somatic hybrids between eggplant and *S. aethiopicum* gr. Gilo. *In*: Proceedings of the 11th Eucarpia Meeting on Genetics and Breeding of *Capsicum* and Eggplant, Antalya, Turkey, pp. 117–120.

Rizza, F., G. Mennela, C. Collonnier, D. Sihachakr, V. Kashyap, M.V. Rajam, Prestera and G.L. Rotino. 2002. Androgenic dihaploids from somatic hybrids between *Solanum melongena* and *S. aethiopicum* group gilo as a source of resistance to *Fusarium oxysporium* F. sp. *melongenae*. *Plant Cell Reports*, **20**:1022–1032.

Rokka, V.M., L. Pietila and E. Pehu. 1996. Enhanced production of dihaploid lines via anther culture of tetraploid potato (*Solanum tuberosum* L. ssp. *tuberosum*) clones. *Am. Potato J.*, **73**:1–12.

Rokka, V.M., C.A. Ishimaru, N.L.V. Lapitan and E. Pehu. 1998. Production of androgenic dihaploid lines of the disomic tetraploid potato species *Solanum acaule* ssp. *acaule*. *Plant Cell Reports*, **18(1/2)**:89–93.

Rotino, G.L. 1996. Haploidy in egg plant. In: *In vitro* production of haploids in higher plants, Vol. 3, S.M. Jain, S.K. Sopory and R.E. Veillux (eds.), Kluwer Academic Publishers, Amsterdam, pp. 115–124.

Rotino, G.L., A. Falavigna and F. Restaino. 1987. Production of anther-derived plantlets of egg plant. *Capsicum Newsletter*, **6**:89–90.

Rotino, G.L., M. Schiavi, E. Vicini and A. Falavigna. 1991. Variation among androgenic and embryogenic lines of eggplant (*Solanum melongena* L.). *J Genet. Breeding*, **45**:141–146.

Rotino, G.L., G. Mennella, F. Fusari, G. Vitelli, M.G. Tacconi, A. D' Alessandro and N. Acciarri. 2001. Towards introgression of resistance to *Fusarium oxysporum* F. sp. *melongena* from *Solanum integrifolium* into eggplant. *In*: Proceedings of the 11th Eucarpia Meeting on Genetics and Breeding of *Capsicum* and Eggplant, Antalya, Turkey, pp. 303–307.

Schery, R.W. 1952. *Plants for Man*, Prentice-Hall Inc., New York.

Simon, P.W. and S.J. Peloquin. 1977. The influence of paternal species on the origin of callus in anther culture of *Solanum* hybrids. *Theor. Appl. Genet.*, **50**:53–56.

Singsit, C. and R.E. Veilleux 1989. Affecting variability in anther culture and in intra- and interspecific transmission of androgenetic competence in diploid potato species. *Euphytica*, **43**:105–112.

Sinha, S., K.K. Jha and R.P. Ray. 1978. Segmentation pattern of pollen in anther cultures of *Solanum surattense*, *Luffa cylindrica* and *Luffa echinata*. *Phytomorphology*, **28**:43–49.

Sinha, S., K.K. Jha and R.P. Ray. 1979. Callus formation and shoot bud differentiation in anther culture of *Solanum surattense*. *Can J. Bot.*, **57**:2524–2527.

Sirasak, T. and R.E. Veilleux. 1998. Indifference of potato anther culture to colchicine and genetic similarity among anther-derived monoploid regenerants determined by RAPD analysis. *Plant Cell Tiss. Org. Cult.*, **53(1)**:49–58.

Snider, K.T. and R.E. Veilleux. 1994. Factors conversion of androgenetic embryos of *Solanum phureja*. *Plant Cell Tiss. Org. Cult.*, **36**:345–354.

Sonnino, A.S., M. Tanaka, M. Iwanaga and L. Schilde-Rentschler. 1989. Genetic control of embryo formation in anther culture of diploid potatoes. *Plant Cell Rep.*, **8**:105–107.

Sopory, S.K. 1977. Development of embryoids in cultivated pollen culture of dihaploid *Solanum tuberosum*. *Z. Pflenzen physiol.* **84**:453–457.

Sopory, S.K. and B.H. Tan. 1979. Regeneration and cytological studies of anther and pollen calli of dihaploid *Solanum tuberosum*. *Z. Pflenzen zuecht.* **82**:31–35.

Sopory, S.K., E. Jacobsen and G. Wengel. 1978. Production of monoploid embryoids and plantlets in cultured anther of *Solanum tuberosum*. *Plant Sci. Lett.*, **12**:47–54.

Tanksley, S.D., M.W. Ganal, J.P. Prince, M.C. Devicente, M.W. Bonierbale, P. Broun, T.M. Fulton, J.J. Giovannoni, S. Grandillo, G.B. Martin, R. Messeguer, J.C. Miller, L. Miller, A.H. Paterson, O. Pineda, M.S. Roder, R.A. Wing, W. Wu and N.D. Young. 1992. High-density molecular linkage maps of the tomato and potato genomes. *Genetics*, **132**:1141–1160.

Ugandhar, T. 2002. Tissue culture studies in *Solanum surattense* Burm. f. Ph.D. Thesis submitted to Kakatiya University, Warangal, A.P., India.

Uhrig, H. and F. Salamini. 1987. Dihaploid plant production from 4 × genotypes of potato by the use of efficient anther plants producing tetraploid strains (4x EAPR clones) — proposal of a breeding methodology. *Plant Breed.* **98**:228–235.

Uhrig, H. and G. Wenzel. 1987. Breeding for virus and nemadote resitance in potato through microspore culture. In: Biotechnology in Agriculture and Forestry Vol. III. Y.P.S. Bajaj (ed.), Springer-Verlag, Berlin, Germany, pp. 346–357.

Waari, S. 1996. The potential of using dihaploid/haploid genotypes in breeding potato by somatic hybridization. *In*: *In vitro* — production in Higher plants S.M. Jain, S.K. Sopory and R.E. Veillaux (eds.), Vol. 3, 321–338, Kluwer Academic Publishers, Dordrecht, The Netherlands.

Weatherhead, M.A. and G.G. Henshaw. 1979. The production of homozygous diploid plants of *Solanum verrucosum* by tissue culture technique. Euphytica, **29**:765–768.

Wenzel, G., O. Scheider, T. Przewozny and S.K. Sopory. 1979. Comparison of single cell culture derived *Solanum tuberosum* L. plants and a model for their application in breeding programs. *Theor. Appl. Genet.*, **55**:49–55.

Zenkleter, M. 1973. *In vitro* development of embryos and seedlings of pollen grains of *Solanum dulcamara*. Z. Pflanzenphysiol., **69**:189–192.

In vitro Propagation of *Syzygium alternifolium* (Wight) Walp.

An Endemic Medicinal Tree of Tirumala Hills

N. Ramamurthy and *N. Savithramma**

ABSTRACT

A rapid and efficient protocol for regenerating shoots from seed of *S. alternifolium* was developed. Seeds were inoculated on MS medium containing different combinations and concentrations of growth hormones. MS medium supplemented with BAP (2.0 mg/l) along with 0.2 (mg/l) NAA induced more number of shoots than reported earlier. *In vitro* shoots were rooted on ½ strength MS medium supplemented with IBA (0.5 mg/l). The regenerated plantlets were later acclimatized in the green house.

Introduction

Syzygium alternifolium Walp. is an endemic and medicinal plant species of Tirumala hills which are a part of Eastern Ghats of India, situated between 79°19′ to 79°23′ E longitude and 13°37′ to 13°43′ N latitude. The plant is found in open areas of rocky terrane on the hills at 2000–2500 feet altitude where the soil is red and calcarious shales. This area is one of the 'Hot Spots' of endemism and 11th microendemic centre among the 40 found in India. The vegetation of this area is excessively exploited for medicinal, timber and fuel purposes. Due to changing ecoclimtic conditions, pollution, summer fires, indiscriminate cuttings of plants by tribals (Yanadhis, Sugalis, Nakkalas) and people of surrounding village, this species has become rare and its survival has been endangered.

* Dept. of Botany, S.V. University, Tirupati 517 502, India.

Syzygium alternifolium Walp. (Myrtaceae) is a small tree. The flowering is in the month of April and May and fruiting is in June and July (Plate Ia). The plant is pollinated by insects and propagated by seeds.It is reduced in its population density and natural spread and became rare as it is over exploited for its profuse medicinal properties.

Fruit pulp and alcoholic extraction of seeds possess antidiabetic properties. The juice of fresh leaves and pulp of the tender shoots are used to treat bacillary dysentry (Savithramma and Sulochana, 1997). Extract of stem bark possess antiseptic properties. The juice of fruits used to cure stomach-ache and its ulcers. External application of fruit pulp reduced rheumatic pains (Madhava Chetty and Rao, 1990). The fruits are very nutritious and edible. The ripen fruits are also used in making squashes, jellies and vinegar (Rao and Raja Reddy, 1983). Leaves are the best source for economically important compounds such as sitosterol (+) pintol, sideroxylin (Rajasekhar Rao *et al.*, 1989) and sizalterin (Rajasekhar Rao and Srinivasa Rao, 1991). *S. alternifolium* yields durable timber, which is used for preparation of craftes, scantlings, beams, poles and agricultural implements (The wealth of India, 1976).

Natural seed production of this species is erratic and the available seeds possess short viability, disease, and pest problems (Plate Ib). Older trees are difficult to propagate through cuttings. Sexual as well as vegetative propagation methods of *S. alternifolium* are facing many problems and restrict its multiplication on large scale. In addition, factors such as deforestation activities, dam construction, monoculture on the hills, microclimatic changes and pollution have already affected the survival of this species (Madhava Chetty and Rao, 1990). Hence, the present study was undertaken to develop an efficient protocol for production of plantlets through micropropagation.

Materials and Methods

The seeds of *S. alternifolium* were collected from Tirumala hills of Chittoor district, Andhra Pradesh, India in the month of April–June. Seeds were treated with 50% H_2SO_4 to soften the hard seed coat. The seeds were thoroughly washed in running tap water. They were surface sterilized by dipping them in 0.1% $HgCl_2$ for 10 min and finally given 4–5 rinses in sterile distilled water. The soaked seeds were then inoculated aseptically on Murashige and Skoog (MS) medium (1962), Gamborg's (B_5) medium (1968) and White's medium (1954) fortified with different combinations and concentrations of auxins (IBA and NAA) and cytokinins (BAP, Kinetin, 2-ip) containing 3% (w/v) sucrose and 0.8%(w/v) agar. Cultures were incubated for four weeks at 25 ± 2°C under cool fluorescent light (3000 lux) with 16 h photoperiod. All the experiments were repeated thrice with

Healthy fruits

Seeds affected by seed borers

Single shoot from seed

Multiple shoot buds from seed

Elongation of multiple shoots

Induction of rooting

PLATE I

10 replicates at each time. Individual shoots of optimal size, excised from different shoots were transferred to rooting medium containing (0–1.5 mg/l) IBA and NAA alone or in combinations. Rooted microshoots were washed free of medium and placed in pots filled with potting medium mix (1 soil:1 vermiculite v/v). Pots were covered with polythene bags for two weeks and watered initially with half strength MS basal medium for a week and later with tap water. The rooted shoots were kept under shade for two weeks before transferring to the soil.

Results and Discussion

In the present study an attempt has been made to standardize efficient protocol for regeneration of *S. alternifolium* an important endemic and medicinal plant.

The results of the preliminary experiments for screening suitable media for germination of seed and production of multiple shoots indicated that frequency of shoot regeneration, maximum number of shoots and shoot elongation from seed were produced on MS medium than the Gamborg's (B_5) and White's media. The emergence of the radicle was observed one week after inoculated and plume formation was noticed by 10th day, when the seeds were inoculated on MS medium than water agar medium. The seeds were allowed to grow for two weeks.

Experiments were carried out to examine the effect of various growth hormones like cytokinins (BAP, Kinetin and 2-ip) and in combination with auxins (IBA and NAA) for multiple shoot production. MS medium without growth hormones showed only one single shoot from seed (Plate Ic). The observations revealed that BAP with low concentrations of NAA proved better in terms of frequency of shoot regeneration, number of shoots/seed and shoot elongation (Table 1). A high concentration of BAP in combination with auxins inhibited shoot elongation and resulted in dwarf shoots with reduced leaves. Kinetin and 2-ip induced abnormal shoots, which hindered the shoot elongation after 5–6 weeks. The experiments revealed that 2.0 mg/l BAP with 0.5 mg/l NAA was the best for the frequency of shoot regeneration (97.5%), number of shoots/seed (18.2), (Plate Id) and shoot elongation (6.9 cm) (Plate Ie). The enhancement in number of shoots per explant tissue has also been reported in many plant tissue culture systems containing cytokinins particularly BAP as well as a cell division growth hormones (Sharma *et al.*, 1997; Nalwade and Shitole, 2004; Pareek and Kothari, 2002)

The shoots thus developed were excised and subjected to rooting on half strength MS medium supplemented with various concentrations (0–1.5 mg/l) of auxins (IBA and NAA). Among all the auxins tested for rooting results obtained with IBA (0.5 mg/l) were most satisfactory

Table 1: Effect of various plant growth regulators on shoot regeneration from seed

PGR					*Frequency of shoot regeneration (%)*	*No. of shoots/ seed*	*Shoot length (cm)*
BAP	*Kinetin*	*2-ip*	*IBA*	*NAA*			
0.5	–	–	0.2	–	40.5^{o}	3.9^{no}	5.8^{efg}
0.5	–	–	–	0.2	38.2^{p}	2.6^{p}	5.0^{ij}
1.0	–	–	0.2	–	87.2^{cd}	12.9^{cd}	6.0^{e}
1.0	–	–	–	0.2	76.0^{h}	8.6^{hi}	5.4^{h}
1.5	–	–	0.2	–	61.9^{lm}	4.1^{n}	5.0^{ij}
1.5	–	–	–	0.2	73.5^{j}	8.9^{h}	6.2^{bcd}
2.0	–	–	0.2	–	97.5^{a}	18.2^{a}	6.9^{a}
2.0	–	–	–	0.2	92.2^{b}	15.6^{b}	6.3^{bc}
–	1.0	–	0.2	–	50.5^{n}	4.8^{lm}	5.9^{ef}
–	1.0	–	–	0.2	67.0^{k}	7.2^{k}	5.4^{h}
–	1.5	–	0.2	–	80.4^{f}	11.8^{e}	6.0^{e}
–	1.5	–	–	0.2	87.6^{c}	13.3^{c}	6.5^{b}
–	–	1.0	0.2	–	75.8^{hi}	8.4^{hij}	4.8^{k}
–	–	1.0	–	0.2	62.4^{l}	5.2^{l}	5.1^{l}
–	–	1.5	0.2	–	79.9^{fg}	10.1^{f}	4.3^{m}
–	–	1.5	–	0.2	81.4^{e}	9.7^{fg}	4.6^{kl}

Values are represented above are the mean of 20 replicates
Mean values followed by the same letter are not significantly different at $P \leq 0.05$.
(Students — Newmen — Keuls Multiple Comparison Test).

(Plate If). Similar findings were also reported by several workers in different plant species (Èmmanuel *et al.*, 2000; Jagannatha *et al.*, 2001; Singh and Sudarshana, 2003). However NAA used for rooting purpose induced callusing at the base of the shoots. The regenerates were transferred to pots for hardening. Thus complete plantlets were developed within 4–5 weeks.

Acknowledgements

The authors are thankful to DST, New Delhi, for providing financial support during the work.

References

Gamborg, O.L. and L.R. Wetter. 1975. Plant tissue culture methods. Proire Regional Laboratory, Saskatoon, Canada.
Jagannatha, J., T.H. Ashok and Satyanarayana. 2001. *J. Plant. Biol.*, **28(1)**:99–103.
Emmanuel, S., S. Ignacimuthu and K. Kathiravan. 2000. *Phytomorphol*, **50(2)**:195–200.
Madhava Chetty K. and K.N. Rao. 1990. *Vegetos*, **3(1)**:12–15.
Murashige, T. and F. Skoog. 1962. *Physiol. Plant.*, **15**:473–497.
Nalwada, A.R. and M.G. Shitole. 2004. *Ind. J. Plant Physiol.*, **9(2)**:145–157.
Pareek, A. and S.L. Kothari. 2002. *J. Ind. Bot. Soc.*, **81**:339–343.
Rajasekhar Rao, J. and R. Srinivasa Rao. 1991. *Ind. J. Chem.*, **30**:66.
Rajasekhar Rao, J., R. Srinivasa Rao, P. Elahisaheb and G. Subramanyam. 1989. *Ind. J. Chem.*, **28**:199.

Rao, K.N. and K. Raja Reddy. 1983. Threatended plants of Tirupati and its environs in "An assessment of threatened plants of India" by S.K. Jain and R.R. Rao. BSI, Howrah. pp. 167–168.

Savithramma, N. and Ch. Sulochana. 1997. *Fitoterapia*, **69(3)**:235–254.

Sharma, D., L.K. Pareek and N. Chandra. 1997. *J. Ind. Bot. Soc.*, **76**:207–210.

Singh, K. and M.S. Sudarshana. 2003. *Ind. J. Plant Phisiol.*, **8(2)**:125–128.

The Wealth of India 1976. Raw materials Vol. X, Publication and information Directorate, CSIR, New Delhi, India, p. 93.

White, P.R. 1954. Tissue culture media. *In*: E.P. George and P.D. Sherrington (eds.), Plant propagation by tissue culture. Exegetics Ltd. England, pp. 263–280.

Plant Regeneration from Root Culture of *Embelia ribes* Burm.

K. Shankarmurthy, V. Krishna and *B.A. Rahiman**

ABSTRACT

Micropropagation protocol was established for a threatened medicinal shrub *Embelia ribes* Burm. (Myrsinaceae). Root segments proliferate into luxuriant mass of callus on Murashige and Skoog's (MS) medium supplemented with 1 mg/l 2,4-D and 0.2 mg/l Kn. The interaction of BAP and NAA at the range of 1 to 2 mg/l and 0.1 to 0.5 mg/l respectively, provoked the root calli to differentiate into multiple shoots and roots in a single phase culture. Highest number of shoot buds (60 ± 1.24 per callus) and regeneration of optimal number of root intact shoots (52 ± 1.02 per callus) were noticed at the concentration of 1.5 mg/l BAP and 0.3 mg/l NAA. 95% of the hardened regenerants were well acclimatized to the natural condition and there were no phenotypic differences observed between the *in vitro* regenerated and *in vivo* plants.

Abbreviations: MS-Murashige and Skoog, 2,4-D — 2,4-Dichloro-phenoxy acetic acid, BAP — 6-Benzyl aminopurine, Kn — 6-Furfuryl aminopurine, NAA — α-Naphthalene acetic acid.

Introduction

Embelia ribes Burm. (Myrsinaceae) is a large woody shrub sparsely distributed in the Western Ghats of India (Sharma *et al.*, 2002). In Indian system of medicine Ayurveda the whole plant is used for treating inflammation, rheumatism and fever (Kapoor *et al.*, 1983). Roots are astringent, stomachic, and used in colic flatulence and dyspepsia and leaves are demulcent, depurative and useful in pruritus, sore throat, aphthae, ulcers of mouth, indolecent skin diseases and leprosy (Sharma *et al.*, 2002). The fruits bitter

* Department of Biotechnology, Kuvempu University, Shankaraghatta-577 451, Karnataka, India. E-mail: krishnabiotech2003@yahoo.co.in

in taste, good appetizer; cures tumors, ascites, bronchitis, jaundice and mental disorders and decoctions of the root twice or thrice daily proved to be a very effective medicine during the recent influenza epidemic (Kirthikar and Basu, 1987). Seeds are used as antibiotic, anthelmintic, antituberculosis, alterative and stimulative (Guhabakshi *et al.*, 2001).

Phytochemically fruits contain a quinone derivative alkaloid embelin (3-Undecyl 2,5-dihydroxy, 1,4-benzoquinone), christembine and fatty ingradients (Tyagi *et al.*, 1978 ; Johri *et al.*, 1990). A volatile oil vilangin was also isolated from the dry berries (Rao and Venkateshwaralu, 1961). The effectiveness of the natural embelin isolated from berries used as grain protectants for wheat against the stored-product insect pests, *Sitophilus oryzae, Rhyzopertha dominica* and *Ephestia cautella* (Chander and Ahmed, 1989).

The biological activities of this species have been screened for antifertility (Arora *et al.*, 1996), Anti helmentic (Wanjari, 2000), a medicine for influenza (Menon, 1919) anti bacterial activity (Narang, 1960), urinary tract infections (Tripathi, 1992), anti spermatogenic effect (Gupta *et al.*, 1989) and dental caries prevention (Namba *et al.*, 1985).

Traditional medical practitioners residing at Lakkinakoppa forest range used the leaves of this species to cure critical jaundised condition. The ethnomedical literature also quoted the hepatoprotective properties of leaves of *Embelia ribes* (Keshavamurthy, 1994; Manjunath, 2004).

In recent years, there has been an awakening all over the world for use of organic medicaments in place of synthetic and antibiotics and there has been consistent growth in demand of many of these plant-based drugs and several plant products from diverse species. This has given rise to large-scale collection of several medicinal plants and many of them are on the verge of extinction (Pandey and Bisaria, 1998). Community based *ex vitro* cultivation of rare and endangered medicinal plants is acknowledged as a possible solution to meet raw materials for market demand and to reduce *in situ* harvesting pressure. This paper reports a standardized protocol for the rapid regeneration of plantlets through root callus culture.

Materials and Methods

The growing root tip was used as an explant. The root explants were repeatedly washed with running tap water for 5 min to remove the soil particles. Then they were treated with the mild detergent tween-20 for 2 min and again washed with running tap water for 5–10 min followed by rinsing with the distilled water 4 to 5 times. The washed root segments were surface sterilized with 0.1 % (w/v) aqueous mercuric chloride ($HgCl_2$) solution for 3 to 5 min and finally washed with sterile distilled water for 6 to 8 times under laminar air hood, dissected into 3–4 cm long proximal.

The root segments were blotted with a sterilized blotting paper and carefully inoculated horizontally on MS (Murashige and Skoog's, 1962) medium solidified with agar. The medium was supplemented with cytokinin (BAP and Kn) or auxin (NAA, 2,4-D and IBA) alone or at different combination and concentration.

The growth regulators 2,4-D, NAA, , IBA, (Indole-3-butyric acid), BAP and Kn were tested individually at the range of 1 to 5 mg/l of each. A better callogenic response was noticed on 2,4-D and Kn containing media. So the combinations of these hormones were tested at the range of 1.0 to 3.0 mg/l and 0.1 to 0.5 mg/l respectively. The shoot and root differentiation media consisted with a range of 1 to 2.0 mg/l BAP and 0.1 to 0.5 mg/l NAA. The pH of the medium was adjusted between 5.6 to 5.8 prior to autoclaving. About 50 ml of molten medium was dispensed into sterilized culture bottles and autoclaved at 121°C at 15-PSI (1.06 kg/cm^2) pressure for 15 to 40 min. Each culture bottle was inoculated with five explants and incubated at cool white fluorescent tubes (Philips, India; 1000 Lux) and 65 to 70% relative humidity.

The percentage of survival and responses were recovered at the end of nine weeks of culture. The plantlets with well developed roots and fully expanded leaves were transferred to pots containing a mixture of sterilized soil and litter compost (2:1). Plantlets were acclimatized inside transparent polyethylene boxes and maintained in a growth-room for two weeks under controlled conditions (12-h photoperiod, 25 ± 2°C), and were watered with 1/10th strength of MS salt solution and then gradually exposed to reduce relative humidity by progressively opening the plastic covers over a period of two weeks. Finally, the plants were transferred to the medicinal garden.

Each treatment consisted of 10 replicates and a complete randomized design was used in all experiments. The analysis of variance and mean separation were carried out using Duncan Multiple range test (Duncan, 1955) and statistical significance was determined at 5% level.

Results and Discussion

Multiple shoot induction through *in vitro* culture was carried out using root explants on MS medium fortified with BAP ranging from 1 to 2 mg/l BAP and 0.1 to 0.5 mg/l NAA. Effect of these growth regulators on shoot induction is depicted in Table 1. The root tip explants were cultured on MS medium supplemented with auxins (NAA, 2,4-D and IBA) and cytokinins (BAP and Kn) either alone or in combination to explore their morphogenic potentialities. After 15 days of incubation only a meager callusing was observed on to the media supplemented with NAA, IBA and BAP. Initiation of callus was noticed only on the media supplemented with 0.1 to

Induction of callus from the root segments on MS + 1 mg/l 2,4-D and 0.2 mg/l Kn.

30 days old root callus showing differentiation of shoot buds from the root callus on MS + 1.5 mg/l BAP and 0.3 mg/l NAA

PLATE I

Multiple shoot induction and rhizogenesis from the root callus culture.

60 days old culture showing root intact plantlets with prominent leaves.

PLATE II

Soil transferred root calli regenerants at culture condition.

30 days old potted root callus regenerants.

PLATE III

Table 1: Effect of BAP and NAA on shoot bud induction and regeneration of plantlets through Root callus culture of *Embelia ribes*

Plant Growth regulators mg/l		*Number of shoot buds/callus*	*Number of plantlets recovered/callus*
BAP	*NAA*	*Mean ± SD*	*Mean ± SD*
1.0	0.1	8.2 ± 0.78	6.1 ± 0.73
1.0	0.2	16.1 ± 0.87	14.6 ± 0.69
1.0	0.3	20.1 ± 1.10	19.6 ± 0.96
1.0	0.4	25.2 ± 1.23	24.3 ± 1.16
1.0	0.5	31.6 ± 1.07	31.0 ± 1.05
1.5	0.1	43.1 ± 0.73	40.6 ± 0.84
1.5	0.2	52.2 ± 1.14	47.1 ± 0.73
1.5	0.3	60.0 ± 1.24	52.0 ± 1.02
1.5	0.4	51.2 ± 1.32	48.2 ± 0.78
1.5	0.5	45.0 ± 0.66	42.0 ± 0.94
2.0	0.1	30.2 ± 1.03	26.8 ± 1.03
2.0	0.2	27.2 ± 1.03	25.1 ± 0.87
2.0	0.3	22.1 ± 0.87	21.1 ± 0.99
2.0	0.4	19.1 ± 0.87	15.0 ± 0.81
2.0	0.5	9.0 ± 0.94	7.0 ± 0.81
F value:		2438	2687

The value of each concentration consisted of ± S.D. of 10 replicates.
The P value 1.75 is significantly different at 0.05%.

3 mg/l 2,4-D or 0.1 to 1.0 mg/l Kn. In the majority of the plant species the synthetic hormone 2,4-D interacted with endogenous hormones of the explants and stimulated the cells to proliferate into callous mass (Razdan, 2003). The combination of 2,4-D with Kn at the range of 0.5 to 1.5 mg/l and 0.1 to 0.3 mg/l respectively induced better proliferation of callus than they were tested alone and it was optimized at the concentration of 1 mg/l 2,4-D and 0.2 mg/l Kn (Plate Ia).

Among the different treatments tested, better shooting response was observed when BAP was used in combination with NAA with a range of 1 to 2 mg/l and 0.1 to 0.5 mg/l respectively (Plate Ib). The interaction of BAP with NAA in provoking the caulogenic potency of the callus have been reported in many of the medicinally important species (Saini and Jaiwal, 2000; Ramulu *et al.*, 2002). After 45 days of culture multiple shoot induction and rhizogenesis from the root callus was observed (Plate IIa). After 60 days of culture the maximum number of shoot buds 60 ± 1.24 shoots per callus was seen in treatments with 1.5 mg/l BAP and 0.3 mg/l NAA (Plate IIb). Roots of certain angiosperms are also endowed with a unique capacity to form shoot buds in nature. The shoots developing on root segments of the same plant are genetically uniform (Sudha, 2000).

At optimal concentration of BAP and NAA (1.5 mg/l and 0.3 mg/l) a mean of 52 ± 1.02 root intact shoots were recovered. The survival rate of root intact shoots regenerated from the root callus was 95%. They were hardened at culture condition for about 30 days (Plate IIIa). The potted plants from the culture condition were transferred to the medicinal plant garden for further studies (Plate IIIb).

This report provides a standardized protocol for the rapid micro-propagation of *Embelia ribes* from root callus culture. Well-rooted shoots can be easily derived from root cultures on BAP and NAA containing medium in a single-phase culture. Chaturvedi and Sharma (1981) suggested that root is an alternative method for clonal propagation and germplasm preservation. This approach offers a means for producing identical hundreds of clonal plants and a strategy to conserve the germplasm as it may be possible to avoid the collection of plant material from the wild resources and thus contribute to the conservation of the valuable medicinal plant species of Western Ghats.

Acknowledgements

The authors are thankful to Registrar, Kuvempu University for providing the financial assistance. We acknowledge the help of Dr. Riaz Mahmood, KU, India.

References

Arora, R.B., N. Ghatak and S.P. Gupta. 1996. Antifertility activity of *Embelia ribes*. *J. Res. Indian Med.* **6(2)**:107–111.

Chander, H. and S.M. Ahmed. 1989. Comparative evaluation of fungicidal quinones and natural embelin against some insect pests of storage. *J. Stored Prod. Res.*, **25(2)**:87–91.

Chaturvedi, H.C. and M. Sharma. 1981.Tissue culture of economically important plants. Proc. COSTED ANBS Symp., A.N. Rao (ed.), Singapore, pp. 31–302.

Duncan, D.B. 1955. Multiple range and multiple 'F' test. Biometrics. **11**:1–42.

Guhabakshi, D.N., P. Sensarma and D.C. Pal. 2001. A Lexicon Medicinal Plants of India. Naya Prakashan, Calcutta, pp. 135–136.

Gupta, S., S.N. Sanyal and U. Kanwar. 1989. Antispermatogenic effect of embelin, a plant benzoquinone, on male albino rats *in vivo* and *iv vitro*. *Contraception*, **39(3)**:307–320.

Johri, R.K., S.K. Dhar, G.S. Pahwa, S.C. Sharma, J.L. Kaul and U. Zutshi. 1990. Toxicity studies with potassium embelate, a new analgesic compound. *Indian J. Exp. Biol.*, **28(3)**: 213–217.

Kapoor, V.K., A.S.Chawla, M. Kumar and P. Kumar. 1983. Anti-inflammatory agent in Indian Laboratories. *Indian drug*, **30**:481–488.

Keshavamurthy, K.R. 1994. Medicinal Plants of Karnataka. Karnataka Forest Dept. Bangalore.

Kirthikar, K.R. and B.D. Basu. 1987. Indian Medicinal Plants. Lalit Mohan Basu, Allahabad.

Manjunath, B.K. 2004. Floristic composition of Davanagere District, Karnataka. Ph.D. Thesis, Kuvempu University, India, 386–388.

Menon, K. 1919. *Embelia ribes* — a medicine for influenza. *Indian Forester*, **45**:210.

Murashige, T. and F. Skoog. 1962. A revised medium for rapid growth and bioassays with tobacco tissue cultures. *Physiol Plant*, **15**:473–497.

Namba, T. 1985. Studies on dental caries prevention by traditional medicines (Part-VII). Screening of Ayurvedic medicines for anti-plaque action. *Shoyakugaku Zasshi*, **39(2)**:146–153.

Narang, G.D. 1960. Preliminary studies on anti-bacterial activity of *Embelia ribes. J. Vet. Anim. Husb. Res.*, **5(1)**:73–77.

Pandey, A.K. and A.K. Bisaria. 1998. Rational utilization of important medicinal plants: A tool for conservation. *Indian Forester.*

Ramulu, D.R., K.S.R. Murthy and T. Pullaiah. 2002. *In vitro* propagation of *Cynanchum callialatum. J. Trop. Med. Plants.*, **3**:233–238.

Rao, C.B. and V. Venkateswarlu. 1961. Vilangin a new constituent of *Embelia ribes. Curr. Sci.*, **30(7)**:250–260.

Razdan, M.K. 2003. Introduction to Plant Tissue Culture. Oxford & IBH Pub., New Delhi, pp. 6–13.

Saini, R. and P.K. Jaiwal. 2000. *In vitro* multiplication of *Peganum harmala* — an important medicnal plant. *Indian J. Exp. Biol.*, **38**:498–503.

Sharma, P.C., M.B. Yelne and T.J. Dennis. 2002. Database on Medicinal Plants used in Ayurveda. Central Council for Research in Ayurveda and Siddha, New Delhi. pp. 5, 479.

Sudha, C.G. 2000. Regeneration of plants from *in vitro* root segment of *Holostemma annulare* (Roxb.) K. Schum., a rare medicinal plant. *Curr. Sci.*, 503–506

Tripathi, P.C., J. Sengupta and J. Mukherjee. 1992. The role of *Embelia ribes* in urinary tract infections. *Int. Sem.-Traditional Med.*, Calcutta. pp. 112–113.

Tyagi, R.K., M.K. Tyagi, H.R. Goyal and K. Sharma. 1978. A Chemical Study on Krimiroga. *J. Res. Ind. Med.*, **3**:130–132.

Wanjari, Y.S., C.R. Jangde, G.B. Srikharde and S.C. Vora. 2000. Efficacy of aqueous extract of *Embelia ribes* seeds against *Ascardia galli* in white leg horn chicks, *Indian Vet. J.*, **77(1)**: pp. 73–75.

Studies on the use of a Simple Medium in Tissue Culture Experiments

D. Gayatri and *G. Rama Gopal**

ABSTRACT

Tissue culture has several practical applications leading to improvement in crop productivity, which are not possible by conventional science. Micropropagation technology has brought about a revolutionary means of plant production, where large number of genetically identical propagules can be quickly produced. Industrial protocols, however, are primarily an extension of laboratory procedures. Obstacles exist which hamper the large-scale commercialization of this technology. The most important of these is high production cost. A study was conducted in order to establish and optimize an *in vitro* culture medium using low cost ingredients like fertilizers as macronutrients, vitamin capsules, myo-inositol tablet, commercial sugar and china grass or sago. *In situ* modifications in the media composition were permitted. To explore its use experiments were carried out using tobacco leaves for direct organogenesis.

Key words: *Micropropagation, simple medium, vitamin capsules, fertilizer macronutrients, commercial sugar, china grass*

Introduction

The technique of micropropagation through tissue culture has been stabilized during the last several decades. *In vitro* propagated plants are increasingly becoming the material of choice because of disease control, uniformity and the possibilities of rapid multiplication of a valuable genotype. Large number of commercial micropropagation ventures came up so far all over the world. While offering significant advantages over conventional propagation, commercial propagation is a process with

* Tissue Culture Laboratory, Department of Botany, Sri Venkateswara University, Tirupati-517502.

numerous production problems, which rely on numerous small containers, semi-solid nutrient media and manual aseptic invasion of plant tissues. In order to increase the efficiency of *in vitro* techniques tissue culturists have made use of purified chemicals and sophisticated physical facilities and environment control equipment. Capital costs as well as running costs are high for such facilities. Resultant cost of production is high and only high value ornamental plant species make utilization of such facility commercially profitable. Tissue culture does not need a high-tech affair. Many non-fussy plants can be easily grown on kitchen style media (Taji, 2002). Cost of components for tissue culture medium is a concern for both research and production laboratories (Zimmerman *et al.*, 1995). The use of alternative media components with less expensive inputs but giving comparable results was studied.

Materials and Methods

Simple Medium — Formulation

Plant tissue culture relies on maintaining plant cells in aseptic conditions on a suitable nutrient medium. A wide range of published media has been described for plant culture. In the present study an attempt in this direction was made to formulate a simple medium to bring down the cost by using locally available cheaper substitutes of the media components (Table 1). Different ingredients, their concentrations and cost of each ingredient used in the preparation of simple medium are given in Table 2.

Applications

To judge the suitability of simple medium routine tissue culture experiments were conducted using this medium. All explants were washed with 5% teepol for 10–15 min before surface sterilization.

Direct Organogenesis

Small bits of tobacco leaves which were sterilized using 70% alcohol (1 min), 0.1% $HgCl_2$ (5 min) were inoculated on the solid fertilizer medium fortified with BAP (1,2 mg l^{-1}) along with NAA (0.1, 0.5, 1 mg l^{-1}). *In situ* modifications of the media composition were permitted to induce direct organogenesis in tobacco leaves. Variations in simple medium were permitted by increasing twice the normal concentration of each individual fertilizer at a time, using natural growth regulators like coconut milk and different concentrations of sago as a gelling agent.

Table 1: List of Ingredients (with weights) used in the preparation of Simple medium

Components	*Amount Used (mg l^{-1})*
Fertilizer Macronutrients	
$(NH_4)_2SO_4$	6.91
Urea	335.6
KNO_3	2026.8
$Ca(NO_3)_2.4H_2O$	1588.0
$NaNO_3$	0.1756
DAP	258.69
$MgCl_2.6H_2O$	300.0
Micronutrients	
H_3BO_3	6.2
$Na_2MoO_4.2H_2O$	0.25
$CoCl_2.6H_2O$	0.025
$CuSO_4.5H_2O$	0.025
$ZnSO_4.7H_2O$	8.6
$MnSO_4.7H_2O$	22.3
KI	0.83
$FeSO_4. 7H_2O$	27.85
$Na_2EDTA. 7H_2O$	37.25
Glycine	2.0
Becosules (Vit-B Capsule)	½ Tablet
Inositol Tablet (500 mg)	½ Tablet
Commercial Sugar	30.0 g
China Grass	8.0 g

Table 2: List and the cost of ingredients of simple medium

Serial No.	*Name of the Ingredient*	*Cost of the Ingredient*	*Amount of Ingredient used (mg l^{-1})*	*Cost for preparing 100 liters of simple medium (in paise)*
1.	Ammonium Sulphate	Rs. 280.00/50Kg bag	691.00	36.8
2.	Urea	Rs. 250.80/50Kg bag	335.60	16.83
3.	DAP	Rs. 485.60/50Kg bag	258.69	25.12
4.	KNO_3	Rs. 95.00/500 g	2026.80	3850.92*
5.	$Ca(NO_3)_2.4H_2O$	Rs. 82.00/500 g	1588.00	2604.32*
6.	$NaNO_3$	Rs. 95.00/500 g	0.1756	0.33*
7.	$MgCl_2.6H_2O$	Rs. 113.00/500 g	300.00	678.00*
8.	Becosules	Rs. 10.14/ 10 tablets	½ capsule	5070.00
9.	Canesugar	Rs. 7.00/500 g	30.00g	4200.00
10.	China grass	Rs. 9.00/ 10 g	8.00g	7200.00
11.	Sago	Rs. 27.00/Kg	25.00g	67.5

Micro, Iron, KI and Glycine sources are same as that of MS. * Laboratory chemicals used.

Results and Discussion

Micropropagation has become a routine procedure, but the high costs involved have prevented laboratories with limited resources from benefiting from tissue culture technology. Research grade chemicals are comparatively free from impurities, but are relatively expensive. Formulating tissue culture media from household materials like fertilizers, vitamin capsule, commercial sugar, pudding agar can be one means of reducing the high cost of production of plants through tissue culture. Simple medium was prepared using small amounts of locally available, water-soluble inorganic fertilizers as macronutrients, which gave ionic concentrations nearer to those present in MS medium. Among the different fertilizers available in the market, some were omitted due to difficulties in their solubilities or due to the difficulties in attaining equivalent ionic concentrations as those in MS medium. Because of the above problems encountered while preparing the medium, calcium ammonium nitrate, single super phosphate, elemental sulphur and other complex fertilizers could not be used during the preparation of the medium.

Most intact plants, tissues and organs take up nitrogen more effectively and grow more rapidly on nutrient solutions containing both nitrate and ammonium ions, than they do on solutions containing just one of these sources, although in most media, reduced nitrogen is present in lower concentration than nitrate. Two important factors to be considered when trying to find media formulations suitable for different plant species and different kinds of cultures are, total concentration of nitrogen in the medium and the ratio of nitrate to ammonium ions. As in MS medium, both forms of nitrogen (ammonia and nitrates) were included in the simple medium and the reduced form of nitrogen was present in lower concentrations than nitrate. Plants are able to absorb urea. Urea can be used as the sole nitrogen source for cultures, but growth is less rapid than when ammonium and nitrate ions are supplied (Kirby *et al.*, 1987). Urea, which is soluble in water, has only amide as a nitrogen source but when dissolved in water amide group changes into ammonia. Nitrate ions are utilized in preference to urea, when both nitrogen sources are available (King, 1977). Urea is able to serve as a reduced nitrogen source during embryogenesis (Durzan, 1987), but has been used in relatively few culture media, and of these, none has been widely adopted (George *et al.*, 1987). Thus in the simple medium prepared, ammonium ion was supplemented in the form of ammonium sulphate, diammonium phosphate and urea where as nitrate was contributed by calcium nitrate, potassium nitrate, and sodium nitrate.

When compared to MS medium (1962) composition, the concentrations of calcium and sulphate were slightly more and chloride ion concentration was less. However, the total ionic concentration of simple medium is slightly more than that of MS medium. Becosules capsule powder, which is soluble in water, contains all B-complex vitamins, including ascorbic acid, were used instead of pure synthetic vitamins. Half inositol tablet of 500 mg was added to supplement the activity of myo-inositol of MS medium. Commercial grade sugar is an obvious replacement for tissue culture grade sucrose at 10% of its price (Kodym and Arias, 2001). 30 grams of cane sugar normally used in kitchen was found suitable.

In India, Ganapathi *et al.* (1995) reported that commercial grade sugar could replace analytical grade sucrose, with no significant change in the frequency of shoot formation in banana. Refined white and unrefined brown sugars in the Philippines were effective for coconut embryo culture (Bonaobra *et al.*, 1994). It is usually unnecessary to use highly purified agar for commercial propagation work, which is fortuitous since the high cost would be disadvantageous (George, 1993). Boxus (1978) reported the successful use of a relatively cheap agar for industrial propagation. China grass (8 gr) is proved as a better gelling agent like that of expensive agar. Micronutrients, potassium iodide, iron and glycine sources were added in the same proportions as that of MS medium constituents since these were used in very minor quantities. Of all the ingredients used in simple medium, china grass is the only expensive ingredient. Cost of different ingredients used in SM was given in Table 2. The present SM (2 Rs./lit) was considered to be less expensive when compared to MS medium (8 Rs./lit).

Direct Organogenesis

Tobacco was an ideal material based on which MS medium has been formulated. So in the present investigation a number of experiments were carried out using tobacco leaves for direct organogenesis. Simple media containing 1 mg l^{-1} BAP and 0.1 mg l^{-1} NAA induced proliferative growth of leafy material and with high mean number of shoots per explant (19.25) (Table 3). In the present study, a number of experiments were conducted on simple fertilizer media by varying the concentrations of different ingredients used in simple medium, to observe the effects of varying concentrations of different ingredients using tobacco leaf bits for direct organogenesis. Effect of different ionic strengths of simple fertilizer medium supplemented with 1 mg l^{-1} BAP and 0.1 mg l^{-1} NAA were studied which revealed that, full strength medium showed satisfactory results both in development of leafy callus and regeneration of multiple shoots (19.25) (Plate Ia) than half and double the strengths of simple media. However, half the strength showed better response than double the strength of the medium. Effect of different fertilizers used in simple medium were

Direct organogenesis (tobacco leaves) on

(a) Full strength SM + 1 mg l^{-1} BAP + 0.1 mg l^{-1} NAA; (b) SM + 691 mg l^{-1} $(NH_4)_2$ SO_4; (c) SM + 300 mg l^{-1} $MgCl_2$. $6H_2O$; (d) SM + 2 times concentration of iron; (e) SM + 2026.8 mg l^{-1} KNO_3; (f, g) SM + 18%, 20% Sago; (i, j, k) SM + + 10%, 20%, 25% CM

PLATE I

Table 3: Effect of different concentrations of BAP and NAA (mg l^{-1}) in simple medium on direct shoot regeneration from tobacco leaf

Observations: 3–4 weeks after inoculation
Gelling Agent: China grass (0.8%)

Serial No.	*BAP (mg l^{-1})*	*NAA (mg l^{-1})*	*Frequency of shoot regeneration*	*Mean number of shoots per explant*
1	1	0.5	76.46 ± 0.83	14.56 ± 0.05
2	2	1	65.9 ± 0.3	7.13 ± 0.2
3	1	0.1	72.16 ± 0.25	19.25 ± 0.25

Values represented in table the mean of 12 replicates.
± standard error.

studied by increasing twice the concentration of each fertilizer in simple medium individually which revealed that ammonium sulphate, magnesium chloride, potassium nitrate and iron yielded better results even when the concentration of these fertilizers were doubled (Plate Ib, c, d, e).

The activities of synthetic plant growth regulators can also be substituted by using natural organic extracts. Coconut milk induced the formation of roots at lower concentrations (10%) and both root and shoot at 20 to 30% concentration of coconut milk, but the induction was poor, when compared to synthetic plant growth regulators and also time consuming (Plate Ih, i, j).

Using sago instead of china grass can further reduce cost of preparation of simple medium. Gelling capacity of sago increased with increase in sago concentration. The physical state of the medium could affect the diffusion of plant growth regulators and nutrients, influencing the micropropagation rate. High prolific growth of tobacco leaves occurred at 20% sago concentration which revealed that the sago was not only a low cost gelling agent but also as efficient as china grass (Plate If, g).

The success of plant tissue culture as a means of plant propagation is greatly influenced by the nature of the culture medium used. The number of media used by a micropropagation laboratory mostly depends on purified chemicals, which are of highly expensive. So an attempt has been made to bring down the cost by using locally available cheaper substitutes of media components. Experimental results proved that this medium is not only a low cost but also as efficient as MS medium. Aim of this work was to alert emerging laboratories to readily available and inexpensive alternatives for micropropagation.

Refrences

Taji, A. 2002. Plant tissue culture for Home Gardeners at Organja.com.

Kodym, A. and F.J.Z. Arias. 2001. Low-cost alternatives for the micropropagation of banana. *Plant Cell Tiss. Org. Cult.,* **66**:67–71.

Bonaobra, Z.S., E.P. Rillo and A.W. Ebert. 1994. Effect of table grade sugars on growth and development of coconut (*Cocos nucifera*) embryos *in vitro*. Presented as poster paper at the 24th International Horticultural Congress, Kyoto International Conference Hall, Kyoto, Japan.

Boxus, Ph. 1978. The production of fruit and vegetable plants by *in vitro* culture. *In*: Actual possibilities and perspectives. 1978. Hughes *et al.* (eds.), pp. 44–58.

Durzan, D.J. 1987. Ammonia: Its analogues, metabolic products and site of action in somatic embryogenesis. 1987b. Bonga and Durzan (eds.), pp. 92–136.

Ganapathi, T.R., J.S.S. Mohan, P. Suprasanna, V.A. Bapat and P.S. Rao. 1995. A low-cost strategy for *in vitro* propagation of banana. *Current Sci.,* **68**:646–649.

George, E.F., D.J.M. Puttock and H.J. George. 1987. Plant Culture Media, Vol. 1, Exegetics Ltd., Westbury, England.

George, E.F. 1993. Plant Propagation by Tissue Culture, Part 1, The Technology, 2nd edition, Exegetics Limited, Edington, Wilts, BA134 QG, England. Chapter 9, The Components of Culture Media. pp. 274–343.

King, P.J. 1977. Studies on the growth in culture of plant cells, growth limitation by nitrate and glucose in a chemostat culture of *Acer pseudoplantanus*. *J. Exp. Bot.,* **28**:142–155.

Kirby, E.G., T. Leustek and M.S. Lee. 1987. Nitrogen nutrition. Bonga and Durzan (eds.). pp. 67–88.

Murashige, T. and F. Skoog. 1962. A revised medium for rapid growth and bio-assays with tobacco tissue cultures. *Physiol. Plant,* **15**:473–497.

Zimmerman, R., S.V. Bhardwaj and I.M. Fordham. 1995. Use of starch-gelled medium for tissue culture of some fruit crops. *Plant Cell Tiss. Org. Cult.,* , **43**:207–213.

Antihyperglycemic Activity of Leaf Callus of *Gymnema sylvestre* R.Br. in Experimental Diabetic Rats

P. Sairam Reddy,[1*] *G. Prasad Babu,*[1] *Badri Kameswara Rao,*[2*] *Ch. Apparao*[2] and *G. Rama Gopal*[1]

ABSTRACT

This study was undertaken to induce callus *in vitro* from *Gymnema sylvestre* (GS) leaf cultures and to investigate the effect of callus ethanolic extract on blood glucose levels of normal and alloxan induced diabetic rats. Ethanolic extract (0.25 g/kg b.w.) of GS leaf callus was given to rats in fasting condition and blood glucose levels were tested after every two hours of treatment. It showed a significant anti-hyperglycemic activity in diabetic rats, where as in normal rats no blood glucose reduction was observed.

Introduction

Diabetes mellitus is a group of metabolic disorders characterized by hyperglycemia. These metabolic disorders include alterations in the carbohydrate, fat and protein metabolisms associated with absolute or relative deficiencies in insulin secretion and/or insulin action. Presently insulin and various oral hypoglycemic agents are available for the treatment of diabetes mellitus with characteristic profiles of side effects(Holman *et al.*, 1991; Williams *et al.*, 1991; Bei *et al.*, 200). Management of diabetes without any side effect is still a challenging task to the medical system. This led to the increasing demand for the natural products with antidiabetic activity with less or no side effects. GS has a good reputation in traditional

[1] Department of Botany, S.V. University, Tirupathi-517502. A.P., India.

* *Present Address*: Department of Biological Sciences, University Delaware, Newark, DE-19711, USA.

[2] Department of Biochemistry, S.V. University, Tirupathi-517502. A.P., India.

medicine as anti-diabetic and is also used as stomachic, diuretic, and remedy for skeletal fractures(Asholkar *et al.*, 1992). It is also called as a wonder drug for diabetes. A series of studies have evaluated the effects of GS leaf extracts on blood sugar levels in experimental animals and human beings (Chattopadhyay *et al.*, 1993; Murakami *et al.*, 1996). Presently a number of commercial and over the counter herbal products are available which contain varying amounts of GS. Investigation of secondary metabolite biosynthesis by plant cell tissue culture has gained momentum world wide in recent years. The present study has been undertaken to develop GS leaf callus cultures and to evaluate its effect on the blood glucose levels of normal and diabetic rats.

Materials and Methods

Animals

Diabetes was induced in male wistar albino rats aged 4 months (b.w ≅ 180 g) by intraperitonial administration of 150 mg/kg b.w. of icecold aqueous alloxan monohydrate (Kameswara Rao *et al.*, 1999; Kameswara Rao and Apparao, 2001). The animals were fed with pellet diet and kept under a constant 12 hrs dark/light cycle. After a fortnight rats with marked hypoglycemia were selected and used for the study.

Leaf Callus

Juvenile healthy GS leaves were collected from wild grown plants in Tirumala hills (A.P., India). Leaves were washed in 1% laboline detergent for 10 minutes followed by repeated washings in running tap water and then distilled water. Surface decontamination was performed by immersing them in 0.05% mercuric chloride (w/v) for three minutes(Sairam Reddy *et al.*, 1998) and then rinsed with sterile distilled water for 4–5 times. Leaf explants (1 cm^2) were inoculated in transparent glass jars (7.5 × 7.5 cm) containing agar gelled (0.8% w/v) MS basal medium(Murashige and Skoog, 1962) containing various concentrations of 2,4-D (4.52, 9.05, 22.62, 45.25 µM) alone or in combination with kinetin (2.32, 4.62, 9.3 µM) or BAP (2.22, 4.4, 9.3 µM). In each experiment 60 explants were cultured and all experiments were repeated thrice. Callus was subcultured at every four weeks and the data like the frequency of callus regeneration, callus fresh weight, dry weight and the callus texture were recorded at every subculture.

Preparation of Ethanolic Extract

Rapidly proliferating green fragile callus yielding highest biomass, was harvested after 4th subculture and 100 g of shade dried callus was extracted with 500 ml absolute ethanol. The extract was concentrated under reduced pressure in Buchi Rotovapor (R114) and stored at 4°C in a refrigerator.

Experimental Design

The rats were divided into 5 groups (six rats in each group). Group I, normal rats; Group II, normal treated rats; Group III, untreated diabetic rats; Group IV, treated diabetic rats and Group V, diabetic rats treated with 0.2 mg/kg b.w. of Glibenclamide. Blood was collected from the overnight fasted rats before treatment for glucose estimation. The group II and IV rats were given a oral dose of 0.25 g/kg b.w of GS leaf callus ethanolic extract (EtGS) suspended in water. Group I and III rats received distilled water as control. All the extracts were fed to the rats by gastric incubation. Blood samples were collected in heparin by tail cutting for blood glucose estimation at 0, 2, 4, 6 and 8hrs after the treatment. The blood glucose levels were estimated by Ames glucometer (USA) using glucostix.

Results and Discussion

The callus initiation was noticed from the midrib region of leaf after 4 weeks of inoculation. Compared to all other growth regulators, 2,4-D (9.05 µM) showed rapid callus initiation although the biomass yield was less. Addition of kinetin (9.3 µM) to the 2,4-D medium produced rapidly proliferating green and fragile callus (Plate 1). Although BAP in combination with NAA produced highly compact greenish callus it did not proliferate further and turned brown during second and third subculture. Callus developed on MS medium fortified with 2,4-D (9.05 µM) in combination with kinetin (9.3 µM) was found to be optimal for further proliferation, biomass yield and maintenance (Table 1). This callus was used for the further experiments.The synergistic effect of kinetin and 2,4-D on callus production has been well demonstrated(Xie and Hong, 1991). It was found that the rapidly multiplying fragile green callus produces maximum amounts of secondary metabolites. The reason might be due to the maintenance of stable physiological activities in such callus even after continuous subcultures(Amerato *et al.*, 1990).

Fasting blood glucose levels of all five groups of rats at 0, 2, 4, 6 and 8 hours after the ingestion of EtGS were given in the Figure 1. The extract showed 15.4% of antihyperglycemic activity at 4th hour and maximum activity at 6th hour (34.4%) in diabetic treated rats. Treatment did not show any decrease in blood glucose levels in normal treated rats, while, the diabetic rats treated with glibenclamide showed a significant decrease in blood glucose levels. This indicates the presence of active principle in the ethanolic extract of GS leaf callus. The results are in agreement with the report of Shanmugasundaram *et al.*(1990). However Chattopadhyaya

Table 1: Effect of Plant growth regulators on callus induction from *Gymnema sylvestre* leaf. Observations were taken 4 weeks after second subculture

Plant Growth Regulators (µM)			*Nature of the Callus*	*Callus Fresh weight/vial*† *(g)*	*Biomass Yield (%)*
Kinetin	*BA*	*2,4 D*			
–	–	4.52	Loose Fragile	5.3 ± 1.4	1.0
–	–	9.05	Loose Fragile	45.6 ± 3.3	0.8
2.32	–	9.05	Light Green Fragile	40.2 ± 4.1	1.2
4.62	–	9.05	Green Fragile	32.1 ± 3.8	2.3
9.30	–	9.05	Green Fragile	34.5 ± 2.8*	4.8*
–	2.22	9.05	Green Fragile	24.3 ± 0.8	2.2
–	4.40	9.05	Green Compact	16.7 ± 1.0	3.1
–	8.90	9.05	Green Compact	11.4 ± 0.5	3.4
–	–	22.62	Loose Fragile	18.2 ± 1.4	0.6
–	–	45.25	Loose Fragile Watery	19.2 ± 2.6	0.3

† Transparent glass jars with 7.5 × 7.5cm dimensions (Diameter × depth)

* Found suitable to produce more amounts of dry callus and for long time maintenance

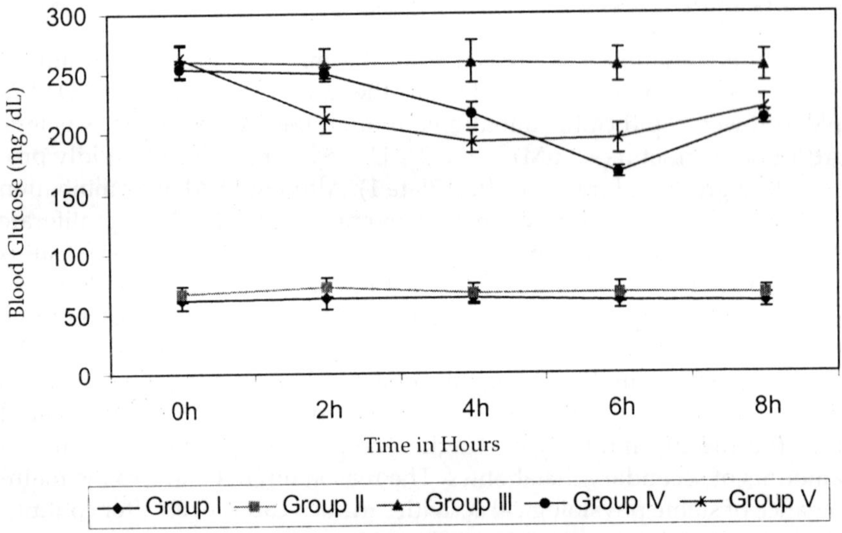

Figure 1: Effect of EtGS on fasting blood glucose levels (mg/dl) in normal and diabetic rats (Mean ± SD)

et al. (1993), showed the blood glucose lowering effect of GS ethanolic extract in normal treated rats also. The antihyperglycemic activity in diabetic rats treated with GS leaf callus extract might be due to the insulin releasing effect from the residual beta cells of pancreas.

A variety of compounds with antidiabetic activity were reported from the leaves of *G. sylvestre* (Miyatake *et al.*, 1993; Ota and Ariyoshi, 1995). The present investigation reports on the production of hypoglycemic

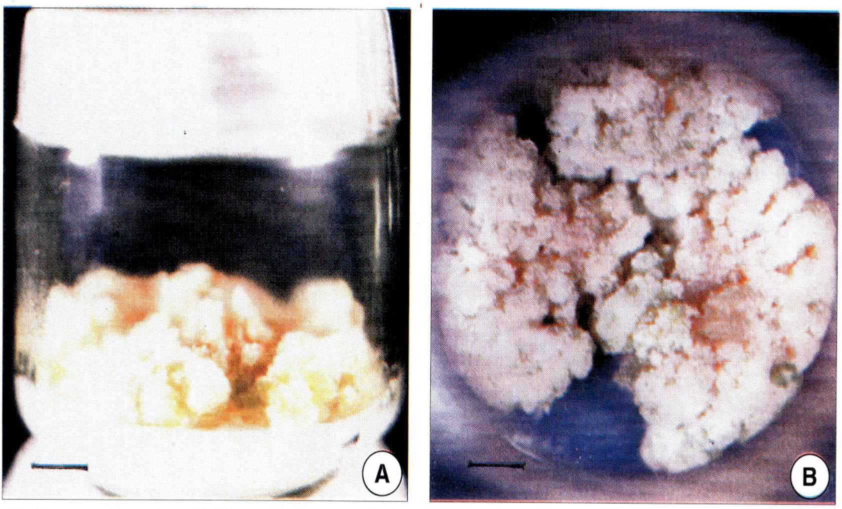

Callus produced on MS medium containing 2,4-D (9.05 mM) + Kinetin (9.3 μM) after 8 weeks of culture.

PLATE I

active principle in *G. sylvestre* leaf callus and might help to produce the precise anti-diabetic active principle at higher quantities. Further studies might be necessary to screen the occurrence of various therapeutically useful compounds and to improve the quality of active principle(s) associated with antidiabetic activity in the leaf callus.

Acknowledgements

SRP and KRB are thankful to UGC, New Delhi for providing financial support in the form of Senior Research Fellowship.

References

Amerato, P.V., D.V. Evans, W.R. Sharp and Y.P.S. Bajaj. 1990. Hand book of plant cell Culture. Vol V. McGraw-Hill Inc. New York.

Asholkar, L.V., K.K. Kakkar and O.J. Chakri. 1992. Second supplement to Glossary of Indian Medicinal plants with active principles. CSIR publications, New Delhi. p. 341.

Bei, B. Zhang and David E. Moller. 2000. *Current Opinion in Chemical Biology*, **4**:461–467.

Chatopadhyay, R.R., C. Medda, S. Das, T.K. Basu and G. Poddar. 1993. Hypoglycemic and antihyperglycemic effect of *Gymnema sylvestre* leaf extract in rats. *Fitoterapia*, **64(5)**:450–454.

Holman, R.R., R.C. Turner, J. Pickup and G. Williams (eds.) 1991. Text book of Diabetes. Blackwell, Oxford, pp. 467–469.

Kameswararao, B. and Ch. Apparao. 2001. Hypoglycemic and antihyperglycemic activity of *Syzygium alternifolium* (Wt.) Walp. Seed extracts in normal and diabetic rats. *Phytomedicine*, **8**:88–93.

Kameswararao, B., M.M. Kesavulu, R. Giri and Ch. Apparao. 1999. Antidiabetic and hypolipidemic effects of *Momordica cymbalaria* Hook. Fruit powder in alloxan induced diabetic rats. *J. Ethnopharmacology*, **67**:103–109.

Miyatake, K., S. Takenaka, T. Fujimoto, G. Kensho, S.P. Upadhyaya, M. Kirihata, I. Ichimoto and Y. Nakano. 1993. Isolation of Conduritol A from *Gymnema sylvestre* and its effects against intestinal glucose absorption in rats. *Biosci. Biotech. Biochem.*, **57(12)**:2184–2185.

Murakami, N., T. Murakami, M. Kadoya, H. Matsuura, N. Yamahara and M. Yoshikawa. 1996. New hypoglycemic constituents in Gymnemic acid from *Gymnema sylvestre. Chem. Phar. Bul.*, **44(2)**:469–471.

Murashige, T. and F. Skoog. 1962. A revised medium for rapid growth and bioassay with tobacco tissue cultures. *Physiol Planta.*, **15**:473–497.

Ota, M. and Y. Ariyoshi. 1995. Location of the disulfide bonds of the sweetness-suppressing polypeptide gurmarin. *Biosci. Biotech. Biochem.*, **59**:156–157.

Sairamreddy, P., G. Rama Gopal and G. Lakshmi Sita. 1998. *In vitro* multiplication of *Gymnema sylvestre* R.Br. and important medicinal plant. *Current Science*, **75**:843–845.

Shanmugasundaram, E.R., K.L. Gopinath, K. Radha Shanmugasundaram and V.M. Rajendran. 1990b. Possible regeneration of the islets of Langerhans in streptozotocin diabetic rats given with *Gymneme sylvestre* leaf extracts. *J. Ethnopharmacol.*, **30(3)**:265–279.

Williams, G., J.C. Pickup and G. Williams (eds.). 1991. Text book of Diabetes. II. Blackwell, Oxford, pp. 977–993.

Xie, D. and Y. Hong. 2001. *In vitro* regeneration of *Acecia mangium* via organogenesis. *Plant Cell Tiss. Org. Cult.*, **66**:167–173.

High Frequency *in vitro* Shoots Proliferation from Nodal Explants of *Bacopa monnieri* (L.) Pennel. by Thidiazuron

Ch. Paramageetham,[1,2] *G. Prasad Babu,*[1*] and *J.V.S. Rao*[1]

ABSTRACT

Highest frequency of adventitious shoots were produced from nodal explants of *Bacopa monnieri* (Brahmi) by a single step procedure in semisolid Murashige and Skoog's (MS) medium supplemented with 0.9 μM TDZ and 4.4 μM BAP. Lower concentrations of TDZ stimulated shoot production. Inhibition of shoot elongation was overcome by the addition of BAP to TDZ containing medium. The shoot proliferation process was open ended, yielding an average of 120 shoots per explant after 6 weeks of culture.

Abbreviations: BAP-Benzyl Aminopurine; TDZ-Thidiazuron; MS-Murashige and Skoog's medium.

Introduction

Bacopa monnieri (Brahmi) is an important medicinal and neutraceutical herb of India, which is gaining commercial importance. Due to over exploitation of this resource from natural habitats, there is an urgent need to propagate this species rapidly. One propagation strategy that could be employed is to use cell and tissue culture technology to clonally multiply the superior genotypes. Furthermore, the combination of an efficient cell or tissue culture regeneration system with recombinant DNA technology would permit the transfer of desired gene.

---

1 Plant Biotechnology laboratory, Dept. of Botany, S.V. University, Tirupati-517 502, India.

2 S.R. & B.G.N.R., Government Degree College for Men, Khammam-507 002. E-mail: prasad_babu17@yahoo.com

Direct multiple shoot production from nodal explants of *Bacopa monnieri*

PLATE I

To date, a variety of explants have been employed to initiate 'Brahmi' cultures including stem, leaf (Srivastava and Rajani, 1999), flower bud (Tejavathi and Shailaja, 1999) and nodal segments (Ali *et al.*, 1996). However successful Brahmi regeneration is time consuming, involving different combinations of growth regulators and more over regeneration frequency is usually low in the above instances.

The goal of the study reported here was to establish rapid clonal propagation protocol from nodal explants using thidiazuron (TDZ). We examined numerous growth regulator regimes including TDZ, a substitute of phenyl urea compound with both cytokinin like and auxin like effects (Visser *et al.*, 1992). Although it is commonly used as a cotton defoliant, TDZ aids in rapid plant regeneration of a number of plant species through organogenesis (Malik and Saxena, 1992). However the effectiveness of TDZ in adventitious shoot formation has not yet been demonstrated with *Bacopa monnieri*. We report here the results of our experiments on shoot proliferation from nodal explants of *B. monnieri* and the effect of thidiazuron (TDZ) and other growth regulators on the process.

Material and Methods

Nodal segments were collected from Botanical garden of S.V.University, Tirupati. They were surface sterilized by dipping them for 1 min. in 70% ethanol followed by 0.1% $HgCl_2$ for 3 min. and finally five rinses with sterile distilled water. The explants were cultured in each test tube which contained 50 ml of the culture medium, either liquid or semisolid. The culture medium consisted of MS salts (Murashige and Skoog, 1962) and vitamins supplemented with 2% sucrose and 0.4% agar in case of solid medium. The pH of the medium was adjusted to 5.8 before autoclaving at 121°C for 20 min. Filter-sterilized plant growth regulators were added to cooled medium which was dispensed in 50 ml aliquots into sterile culture tubes. The cultures were incubated in a culture room at 24 ± 2°C with a 16-h-photoperiod and irradiance of 70 μmol $m^{-2}s^{-1}$ supplied by cool white fluorescent lamps. Shoots produced from these nodal explants were sub-cultured for every 4 weeks.

For experimental work MS basal medium was supplemented with TDZ ranging from 0.4 μM to 4.5 μM either singly or in combination with 4.4 μM BAP or 4.6 μM kinetin to evaluate the TDZ effect on multiple shoot formation from nodal explants in MS semisolid medium. In our study nodal explants were also cultured in the presence of BAP, Kinetin and Zeatin individually at various concentrations on to MS liquid medium for comparing against TDZ in regeneration efficiency.

After one week in culture, multiple shoot buds appeared in several treatments. The developing shoots were transferred on to MS basal liquid medium for a week to induce rooting/to facilitate growth of the plantlets. After 4 weeks the most vigorously growing plantlets were harvested and transferred to pots containing 1:3 sand and soil, later acclimatized under mist for several days. Then they were transferred to 15 cm pots containing farm yard manure. Each treatment contained 10 replicates and the whole experiment was repeated thrice. Data was statistically analysed.

Results

Most (>85%) nodal explants formed numerous rosette structures within a week when cultured in semi solid medium containing TDZ. A rosette is a compressed stem with nodes, short internodes and associated with leaves. Continued culture in this medium resulted in the formation of a normal shoot from the apical region of this rosette structures.

Since the dose of TDZ is known to be critical in shoot organogenesis (Lu, 1993; Yusnita *et al.*, 1990), here we compared the response of *Bacopa* nodal segments on various TDZ concentrations either singly or combination with 4.4 μM BAP/4.6μM Kinetin (bud initials began to appear within a week after culture initiation). The effects of TDZ concentrations were significant on adventitious bud and shoot formation (Table 1). An increase in the concentration of TDZ from 0.4 to 0.9 μM increased the bud induction, while TDZ concentrations below 1.3 μM to 4.5 μM decreased bud induction.

Table.1: Effect of TDZ alongwith BAP & Kinetin on Shoot multiplication from nodal explants of *B. monnieri*

BAP	*Kinetin*	*TDZ*	*Frequency of response (%)*	*Mean no of shoots**	*Morphegenic response***
Concentration in μM SNK Test was done to compare all concentrations of TDZ Vs all concentration of BAP and kinetin when used individually		0.4	80	21.6[a]	–
		0.9	85	27.0[b]	–
		1.3	80	24.0[c]	C+
		1.8	75	16.0[d]	C+
		2.2	70	13.2[e]	C++
		4.5	55	5.4[f]	C++
4.4	–	0.9	70	120.0	R
–	4.6	0.9	70	70	R
4.4	4.6	–	–	40.3	–

* Mean values followed by same letters are not significantly different at $P \leq 0.05$ (Student — newman-keuls multiple comparison test).

** C+ – Poor
C++ – Moderate
R – Rhizogenesis

At 0.9 μM TDZ concentration, the addition of 4.6 μM kinetin improved shoot number to an average of 70 per explant, but at 0.9 μM TDZ in combination with 4.4 μM BAP bud induction was most prolific with an average of 120 shoots per explant whereas addition of BAP had little effect on bud induction.

After 5 days in culture, numerous shoot buds were formed on each explant (Plate Ia) and after 2 weeks the number of shoot buds increased and at the end of 3rd week shoot buds were elongated (Plate Ib). At the end of 4th week the shoot number significantly increased to an average of 120 shoots alongwith rhizogenesis. The roots were slender and short (Plate Ic). There was no significant difference in adventitious bud or shoot production among segments from different orientation (data not shown).

All treatments involving BAP, Kinetin and Zeatin produced lower number of shoots than TDZ (Table 2). Among all the three cytokinins BAP proved better at 4.4 μM in inducing an average of 20 shoots per explant without intervention of callus. However higher concentration of BAP ranging from 8.8 μM to 23.1μM and Kinetin 9.2 μM to 23.2 μM decreased the

Table 2: Effect of various cytokinin on Shoot multiplication from nodal explants of *B. monnieri*

BAP	*Kinetin*	*Zeatin*	*Frequency response (%)*	*Mean no. of shoots*	*Other responses*
Concentration in μM					
	Control		20	–	–
2.2	–	–	75	11.05[a]	–
4.4	–	–	80	19.05[b]	–
8.8	–	–	85	20.3[cd]	–
13.3	–	–	80	13.7[d]	C+
17.7	–	–	75	10.4[e]	C++
23.1	–	–	65	8.3[f]	C++
–	2.3	–	70	9.0[a]	–
–	4.6	–	75	13.0[b]	–
–	9.2	–	75	14.1[c]	C+
–	13.9	–	70	9.8[d]	C+
–	18.5	–	65	7.0[ef]	C++
–	23.2	–	60	3.8[f]	C++
–	–	2.2	60	0.7[a]	–
–	–	4.5	65	1.4[b]	–
–	–	9.1	70	5.8[c]	–
–	–	13.6	60	3.5[d]	–
–	–	18.2	50	2.5[e]	–
–	–	22.8	45	1.6[f]	–

* Mean values followed by same letters are not significantly different at $P \leq 0.05$ (Student — newman-keuls multiple comparison test).

** C+ – Poor
C++ – Moderate
R – Rhizogenes

shoot number and increased the callus formation. Whereas Zeatin at all concentrations resulted in abnormal shoots.

Thidiazuron has been reported to inhibit the shoot elongation in many plant species (Lu, 1993). In the case of *Bacopa* shoot elongation was not evident till the end of 3rd week on liquid MS medium containing TDZ alone (Fig. 1) indicating an inhibitory effect of TDZ on shoot elongation when compared with BAP and Kinetin. However, BAP (or) Kinetin resulted in normal length of shoots. Additionally, extended cultures with TDZ resulted in abnormal leaf morphology and occasional necrosis of shoot tips. The problem with shoot elongation was overcome by the addition of BAP/Kinetin to TDZ containing medium. There was a significant difference in shoot elongation when Kinetin/BAP used with TDZ.

Rooting of the in vitro *regenerated shoots*

Although shoot formation was associated with rhizogenesis when used TDZ with BAP/Kinetin, in some instances like cytokinins and TDZ when used individually separate rooting experiment was involved. When IBA 0.4μM and 2.4μM) and NAA (0.5 μM and 2.6 μM) were added individually to the MS full strength liquid medium root initiation was observed in all the treatments including control. Though IBA/NAA induced roots with 90% of frequency but, resulted in unwanted callus formation from the basal portions of the shoots (Table 3). While, rooting percentage was high (95%) in MS full strength liquid basal medium without callus intervention. (Plate Id)

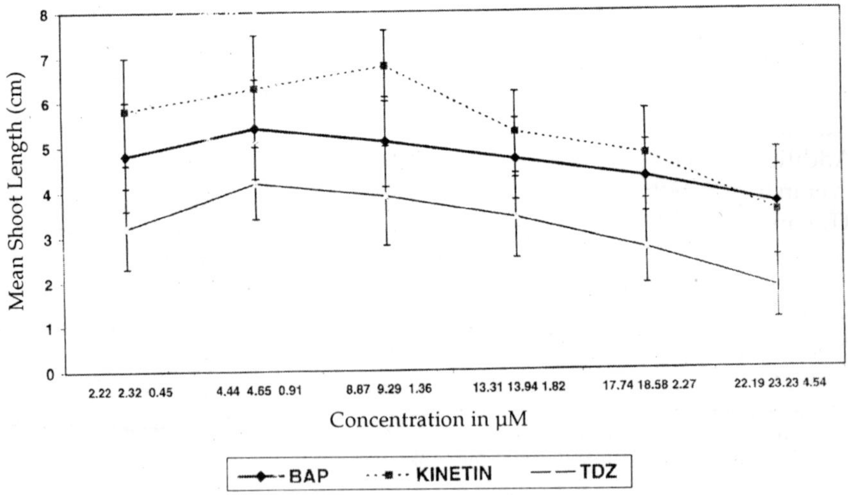

Figure 1: Effect of various Cytokinins when used individually in MS medium on nodal explants of *B. monniera*

Table 3: Effect of NAA and IBA when used individually in MS medium on rooting of isolated shoots of *B. monnieri**

Name of the growth regulator (µM)		*% of rooting*	*Morphogenic Response*
IBA	NAA		
Control		95	Normal long roots
0.49	–	90	roots with basal callusing
2.46	–	90	roots with basal callusing
–	0.54	90	roots with basal callusing
–	2.69	90	roots with basal callusing

* Values represented above are mean of 20 replicates and repeated thrice.

Hardening and Acclimatization

During our study we observed that the nodal explants rooted better when cultured on 0.9 TDZ with Kinetin/BAP. 95% plantlets were successfully acclimatized to the garden conditions (Plate Ie). However loss occurred at the time of exposing the plantlets to the environment by removing bags.

Discussion

Thidiazuron is a substitute of phenyl urea and the biochemical action is not yet completely understood. They are believed to function in the regulation of purine cytokinin metabolism or act directly as cytokinins (Mok *et al.*, 1982; Capelle *et al.*, 1983; Mok and Mok 1985). The biological activity of TDZ is higher than or comparable to that of the most active adenine type cytokinins (Mok *et al.*, 1987; Huetteman and Preece, 1983). In our experiments lower concentrations of TDZ were most effective in higher concentrations of TDZ resulted in decreased shoot number with excessive callus formation and cessation of shoot growth. Higher concentrations TDZ was toxic causing excessive browning and eventual death of the explants. Addition of BAP to TDZ significantly increased the shoot number. However in hypocotyl segments of *Liquidambar styraciflua* addition of 2,4-D with TDZ maximized shoot production. In *B. monnieri* TDZ's inhibitory effect on shoot elongation was overcome by adding BAP/Kinetin as reported by many researchers who found that for elongation other cytokinins or combinations of cytokinins and auxins must be used (Fasolo *et al.*, 1989; Preece and Imel, 1991; Bates *et al.*, 1992). A possible explanation in the context of *B. monnieri* is that BAP continued to stimulate shoot initiation while at the same time reducing the dominance of newly formed shoots on the remaining buds.

It was not uncommon to harvest as many as 120 shoots from a single explant. Such prolific shoot formation cannot be accounted on the basis of precocious axillary branching, since care was taken to prevent axillary buds from making a significant contribution to shoot proliferation. During our study we observed rooting on the same shoot multiplication medium when TDZ and BAP were used. While Sanego *et al.*, (1995) reported inhibitory capacity of rooting by TDZ. Our observation suggests that TDZ may sometimes act like an auxin.

Conclusion

In conclusion our study demonstrates that TDZ promotes high frequency adventitious bud formation and shoot differentiation when used with BAP in *B. monnieri* nodal segments. By using the described protocol high frequency regeneration of certain *B. monnieri* genotypes can be realized with minimal sub culturing, yielding abundant plants within a short time (4–6 weeks) and space which represents a marked improvement over the procedures commonly available. The progeny of the micropropagated plants have shown no obvious variation in normal morphology to date. The described regeneration system may be adopted for the *in vitro* manipulation of *B. monnieri* to obtain super plants of high drug content.

Acknowledgement

The first author is thankful to CSIR for awarding SRF (NET) and financial support during the work.

References

Ali, G., M. Purohit, M.H. Mughal, M. Iqbal and P.S. Srivastava. 1996. A rapid protocol for micropropagation of *Bacopa monnieri* (L.) Wettst. an important medicinal plant. *Plant Tiss. Cult. Biotech.*, **2**:208–211.

Bates, S., J.E. Preece, N.E. Navarrete, J.W. Sambeek van and G.R. Guffney. 1992. Thidiazuron stimulates shoot organogenesis and somatic embryogenesis in white ash *(Fraxinus americana* L.). *Plant Cell Tiss. Org. Cult.*, **31**:21–29.

Cappelle, S.C., D.W.S. Mok, S.C. Kirchner and M.C. Mok. 1983. Effects of thidiazuron on cytokinin autonomy and the metabolism of N^6-(2-isopentyl) [8-C^{14}] in callus tissue of *Phaseolus lunatus* L. *Plant Physiol.*, **73**:796–802.

Fasolo, F., R.H. Zimmerman and I. Fordham. 1989. Adventitious shoot formation on excised leaves of *in vitro* grown shoots of apple cultivars. *Plant Cell Tiss. Org. Cult.*, **16**:75–87.

Huetteman, C.A. and J.E. Preece. 1993. Thidiazuron: a potent cytokinin for woody plant tissue culture. *Plant Cell Tiss. Org. Cult.*, **33**:105–119.

Lu, C-Y. 1993. The use of thidiazuron in tissue culture. *In vitro Dev., Biol.*, **29**:92–96.

Malik, K.A. and P.K. Saxena. 1992 Thidiazuron induces high frequency shoot regeneration in intact seedlings of Pea, Chickpea and Lentil. *Aust. J. Plant Physiol.*, **19**:731–740.

Mok, M.C., D.W.S. Mok, D.J. Armstrong, K. Shudo, Y. Isogai and T. Okamoto. 1982. Cytokinin activity of N-Phenyl-N-1,2,3, thiadiazol-5-Yl Urea (thidiazuron). *Phytochemistry* **21**:1509–1511.

Mok, M.C. and D.W.S. Mok. 1985 The metabolism of [C^{14}] thiadazuron in callus cultures of *Phaseolus lunatus. Physiol. Plant.*, **65**:427–432.

Mok, M.C., D.W.S. Mok, J.E. Turner and C.V. Muser. 1987. Biological and biochemical effects of cytokinin-active phenylurea derivatives in tissue culture system. *Hort. Science*, **22**:1194–1197.

Murashige, T. and F. Skoog. 1962. A revised medium for rapid growth and bioassays with tobacco cultures. *Physiol. Plant*, **15**:473–497.

Preece, J.E. and M.R. Imel. 1991. Plant regeneration from leaf explants of *Rhododendron* 'P.J.M. Hybrids'. *Scientia Hort.*, **48**:159–170.

Sanego, M.H.M., S.J. Murch, T.Y. Slimmon, S. Krishnaraj and M.K. Saxena. 1995. Morphoregulatory role of thidiazuron: morphogenesis of root out growths in thidiazuron-treated geranium. *Plant Cell Rep.*, **15(3)**:205–211.

Srivatsava, N. and M. Rajani. 1999. Multiple shoot regeneration and tissue culture studies on *Bacopa monnieri* (L.) Pennell. *Plant cell Rep.*, **18**:919–923.

Tejavathi, D.H. and K.S. Shailaja. 1999. Regeneration of plants from the cultures of *Bacopa monnieri* (L.) Pennell. *Phytomorphol.*, **49**:447–452.

Visser, C., J.A. Qureshi, R. Gill and P.K. Saxena. 1992. Morphoregulatory role of thidiazuron: substitution of auxin and cytokinin requirement for the induction of somatic embryogenesis in geranium hypocotyl cultures. *Plant Physiol.*, **99**:1704–1707.

Yusnita, S., R.L. Geneve and S.T. Kester. 1990. Micropropagation with flowering eastern redbud (*Cercis canadensis* var'*alba'* L.) *J. Viron Hortic.*, **8**:177–179.

The Role of Phytohormones on Caulogenesis and Morphogenesis in Plant Tissue Culture — Pearl Millet and Green Gram

Karthik Iyer, C. Parthasarathy and *R. Basavaraju**

ABSTRACT

The morphogenic response of young immature inflorescence on MS medium with 2,4-D is good than other treatments and other explants in caulogenesis. Good regeneration was observed in combination of BAP + Kn. The combination of IBA + BAP resulted in good regeneration with better roots. Shoot formation recorded in the treatment of BAP + Kn + NAA and BAP + Kn + IBA is good in Pearl millet.

In Green gram stem explant on MS medium with, 2,4-D yielded good callus than other explants. Leaf explant on 2,4-D + BAP showed better growth of callus than other combinations. Greenish callus was observed in the treatment of NAA + IBA. The results are useful for the study of morphogenesis and somaclonal variation.

Introduction

In India, Pearl millet (*Pennisetum glaucum*(L.)R.Br) is the fourth important cereal food crop and is grown in 10.6 million hectares area with annual production of 8.9 million tonnes. It is widely cultivated in the semi-arid tropics, and is adapted to the dry regions with sandy infertile soils, where rainfall is low and erratic. In Anathapur district Pearl millet is cultivated in 3000 hectares producing 1000 tonnes(1994–95).

Green gram (*Vigna radiata*(L.) Wilczek.), known as mung bean, is indigenous to India and has been in cultivation since prehistoric times. India has about 3 million hectare and produces about one million tonnes of green gram. An attempt has been made to study the effect of different

* Department of Biosciences, Sri Sathya Sai Institute of Higher Learning, Prashantinilayam.

phytohormones individually as well as in combination on the explants from the dicot (green gram) and monocot (pearl millet) in tissue culture.

Materials and Methods

The explants were surface sterilized with mercuric chloride solution (1% w/v) for three minutes seeds in case of green gram and young inflorescence of pearl millet. Then they were washed thoroughly with sterile distilled water for 6–7 times for 3–5 minutes each. Experiments were set with MS medium(Murashige and Skoog, 1962) fortified with phytohormones and cultures were incubated at 25 + 2°C and light intensity of 2000 lux. All aseptic and sterile conditions for the growth of cultures were maintained during the experiment.

Results and Discussions

The young immature inflorescence was found to be better yielder of callus than other explants (stem, leaf) in pearl millet. In the treatment of auxins and cytokinins the response of immature inflorescence was observed to be maximum in 2,4-D (2 mg/l)than others in the caulogenesis. Mishra *et al.* (1993) reported initiation of callus culture in *P.glaucum* on MS medium with 2,4-D. Kulakarni *et al.* (1995) have reported initiation of callus from inflorescence with 2,4-D. Caulogenesis was observed on MS medium with NAA (2 mg/l) + Kn (2mg/l); 2,4-D (2mg/l) + Kn (2 mg/l) combination was better than other treatments. When the callus was sub-cultured on fresh MS medium using the combination of BAP (2mg/l) + Kn (2mg/l); BAP (2mg/l) + IBA (2mg/l); BAP (2mg/l) + Kn (2mg/l) + NAA (2mg/l), BAP (2mg/l) + Kn(2mg/l) + IBA(2mg/l) the regeneration was observed, while BAP (2mg/l) + Kn (2mg/l) + 2,4-D (2mg/l) resulted only in good caulogenesis. MS medium with coconut milk also yielded better results for regeneration. Regeneration started appearing after 20–25 days as short green outgrowth and by 50–60 days regeneration was complete resulting in the formation of plantlets. The triple combination of BAP (2mg/l) + Kn (2mg/l) + NAA (2mg/l) gave good morphogenesis resulting in the shoot formation. BAP (2mg/l) + Kn(2mg/l) + IBA(2mg/l) combination on MS medium resulted in the good organogenesis and rhizogenesis in the formation of good root and shoot system. Kularni *et al.* (1995) have obtained plantlets on MS medium supplemented with IAA, Kn and Adenine. Kalamani and Ramaswamy (1998) have also recorded regeration from inflorescence explants by using N6, White's, L.S. and MS medium. For initiation and induction of callus the auxin 2,4-D (2mg/l) is good for inflorescence explant. For subculture of callus 2,4-D (2mg/l) + Kn (2mg/l) is better than other treatments for proliferation of callus. For regeneration of shoots from the callus the double combination of BAP (2mg/l) + Kn

(2mg/l) is good, whereas the triple combination of BAP (2mg/l) + Kn (2mg/l) + IBA (2mg/l) is better. Coconut milk is best for regeneration and also for caulogenesis.

In case of green gram, stem bits as explant was found to be good as compared to leaves or seeds. Stem explant on MS medium with 2,4-D (6mg/l) gave good caulogenesis when compared to other treatments. Leaf explant on 2,4-D (6mg/l) + BAP (4mg/l) showed better growth of callus with respect to proliferation rate and size of the callus, followed by 2,4-D (8mg/l) + Kn (4mg/l). The callus obtained from the explant was subcultured to study the effect of hormone individually (2,4-D, IBA, IAA, BAP) and in combination (2,4-D + Kn, 2,4D + BAP, IBA + Kn, NAA + BAP, NAA + BAP + Kn and NAA+ BAP + PABA, NAA + BAP + IBA). Extensive greenish calli was observed in the treatment of NAA + BAP + IBA. Light green calli with quite a high proliferation was observed on 2,4-D + Kn, NAA + BAP, IBA + Kn, NAA + BAP + Kn. Singh *et al.* (1980) have studied the effect of different concentrations of hormones in B5 medium on excised shoot apices. They reported that, 2,4-D did not support shoot or root regeneration but there was extensive callus formation. In the present investigation it was observed and recorded that in case of excised shoot apices on MS medium supplemented with 2,4-D supported callus growth. Patel *et al.* (1991), Wali and Siddiqui (2001) have reported the caulogenesis in case of green gram using auxins and cytokinins.

In green gram the calli were light green, compact, cohesive and fresh. In pearl millet they were granular, whitish or whitish green to light green and some what dry. Stem explant was found to be good for caulogenesis in green gram where as in pearl millet young inflorescence is good.

References

Kalamani, A. and N.M. Ramaswamy. 1998. *In vitro* response of genotypes in pearl millet. *Indian Journal of Genetics and Plant Breeding*, **58(2)**:153–158.

Kulkarni, V.M., A.N. Choudhari and B.S. Shah. 1995. *In vitro* flowering in pearl millet (*Pennisetum glaucum* (L.) R.Br.). *Indian Journal of Experimental Biology.*, **33(2)**:142–143.

Mishra, M., N. Das and A.N. Mishra. 1993. Effect of 2,4-D on growth and oxidase activities in callus cultures of pearl millet. *Annals of Plant Physiology*, **7**:1–4.

Murashige, T. and F. Skoog. 1962. A revised medium for rapid growth and bioassays with Tobacco tissue cultures. *Physiol. Plant*, **15**:473–497.

Narayanaswamy, S. 1994. Plant Cell and Tissue Culture. Tata Mc Graw-Hill Publishing Company Limited, New Delhi, pp. 273–287.

Singh, B.D., R.P. Singh, R.B. Singh and R.M. Singh. 1980. Organogenesis in mung (*Vigna radiata* var *aureus*) from macerated shoot tips. Proceedings of National Symposium on Plant Tissue culture, genetic manipulation and somatic hybridization of plant cells. pp. 335–341.

Wali, S.A. and S.A. Siddiqui. 2001. Shoot organogenesis from stem callus in *Vigna radiata* (L.) Wilczek. *Phyiol Mol. Biol.*, **7(2)**:1911–194.

The Role of Phytohormones on Caulogenesis in Plant Tissue Culture — Ground Nut and Mustard

E. Ramulu, P. Sujithkumar and *R. Basavaraju**

ABSTRACT

The response of explants i.e. epicotyl and cotyledon varied with respect to callus induction, its growth and organogenesis. Callus induction was better in 2,4-D and 2,4-D +Kn, than other treatments. Callus from epicotyl when sub-cultured on MS medium with IBA, BAP, IBA + NAA + BAP + Kn resulted in plantlet germination. Callus from cotyledon showed good proliferation on MS medium with NAA + BAP and produced plantlets.

The seedling of mustard 7 day old revealed good amount of callus in the treatment of 2,4-D, Kn NAA, BAP and good rhizogenesis with IBA. Regeneration was recorded in the treatment of NAA + BAP + Kn. Many treatments showed white hair formation on callus.

Introduction

Groundnut (*Arachis hypogaea* L.) is a very important oilseed crop. India occupies first position in terms of acreage and production in the world. In India, 9,164 lakh tonnes of ground nut were produced in 7,572 lakh hectares during 1998–1999. In A.P. ground nut is cultivated on an area of 2.35 million hectares with production of 2.55 million tonnes. India is the leading country in the world regarding the areas under cultivation of Mustard (*Brassica juncea* L.) the oil seed crop. In 1998–1999, the area was 6,598 lakh hectares and production was 5,774 lakh tonnes.

* Department of Biosciences, Sri Sathya Sai Institute of Higher Learning, Prashantinilayam.

Material and Methods

The variety VRI-2 of ground nut, used in the study, was procured from Karnataka Seeds Corporation, Hebbal, Bangalore. Seeds were surface sterilized with mercuric chloride (1% w/v) for three minutes, then they were washed thoroughly in distilled sterile water for 5–6 times for 4–5 minutes each. The seeds (both ground nut and mustard) were allowed to germinate in sterile conical flasks. Hypocotyl, epicotyl and cotyledons of the germinating seedlings of ground nut were used as explants. The seedlings of mustard were used as explants. The explants were inoculated on MS medium supplemented with hormones. The cultures were kept in culture room at 23–25°C temperature, with light intensity of 2000 lux.

Results and Discussions

The responses of explants were found to be varied in terms of callus induction, its growth and organogenesis among the epicotyl, hypocotyl, and cotyledon parts of the ground nut tested. Callus induction was found to be better in 2,4-D (8 mg/l) and 2,4-D (8 mg/l) + Kinetin (4 mg/l) than in the other treatments. The explants from epicotyl, hypocotyl and cotyledon produced good amount of callus and proliferated further on MS medium supplemented with 2,4-D (8 mg/l). Bhuiyan *et al.* (1992) observed that the leaf explant cultured on MS medium with 2,4-D gave good callus but not shoots as in the present study. Callus from cotyledon, callus from hypocotyl and callus from epicotyl showed good proliferation on MS medium fortified with IBA (8 mg/l), BAP (8 mg/l) and 2,4-D (8 mg/l) respectively. The combination of NAA (8 mg/l) + BAP (6 mg/l) + Adenine sulphate (3 mg/l) yielded good amount of callus from the callus of hypocotyl and cotyledon. Tyagi *et al.* (1993), observed that the explants from hypocotyl, epicotyl, leaf and embryonal axes formed callus on MS medium containing various combinations and concentrations of kinetin, BAP, NAA and 2,4-D.

The callus obtained from the epicotyl when subcultured on MS medium fortified with IBA, BAP, IBA + Kn, NAA + BAP, NAA + BAP + Kn, Coconut milk + NAA + BAP resulted in plantlet germination. In 1997, Venkatachalam and Jayabalan observed that callus induction and shoot bud regeneration varied with the genotype, explants and the combinations of the plant growth regulators. NAA + BAP combination produced shoots, rooting was effective with NAA. This present work was in concurrence with the results of the above mentioned authors. The MS medium containing IBA showed plantlet regeneration from epicotyl. Ngo and Quesenberry in 2000, obtained shoot development on medium with low level of IBA. The epicotyl callus showed regeneration on MS medium with NAA and BAP to certain extent. Eapen and George (1993) reported best results in terms of the number of shoots produced per culture in the treatment of NAA and BAP but with the embryonal shoot tip. The callus from the cotyledons showed good proliferation on MS medium fortified with NAA and BAP, and yielded plantlets, similar observation was recorded by Radhakrishnan *et al.* (2000).

The epicotyl callus and hypocotyl callus became friable on the subcultures earlier than the cotyledon callus. The cotyledon callus showed good shoot formation in NAA treatment, while the epicotyl callus showed plantlet regeneration in NAA and BAP combination. White protuberances were observed on the epicotyl and cotyledon callus in 2,4-D treatment. It was generally observed that the rate of callus growth reduced when the plantlet or root formation had started.

The seedling of mustard 7 day old, yielded good amount of callus in the individual treatment of 2,4-D (2 mg/l); Kn (2 mg/l); NAA (2 mg/l); BAP (2 mg/l); and IBA (2 mg/l). The treatment of MS with IBA gave good rhizogenesis. The combination of 2,4-D (2 mg/l) + Kn (2 mg/l) also produced good proliferation of callus followed by NAA (2mg/l) + BAP (2 mg/l), while IBA(2 mg/l) + Kn (2 mg/l) and NAA (2 mg/l) + BAP (2 mg/l) produced limited callus. George and Rao (1978) have reported limited amount of callus on MS medium containing NAA and IBA. The MS medium with BAP first produced callus, later on this callus became green and granular. The combination of NAA and BAP both auxin treatment yielded proliferation of callus, which was green and granular. Among various treatments, coconut milk + GA_3 produced green regions faster than other treatment. Sharma *et al.* (1991) have reported that GA_3 inhibited growth and differentiation of shoot from the green callus. In the present work, GA_3 would have inhibited differentiation of shoot from the green callus. The callus obtained from 2,4-D was subcultured on fresh MS medium with BAP + NAA; BAP + NAA + Kn; BAP + NAA + GA_3; BAP + NAA + AS; BAP + NAA + PABA; BAP + NAA+ Coconut milk; out of these only in NAA+ BAP + Kn the regeneration was observed. The callus subcultured on MS medium fortified with NAA + BAP+ Kn showed better amount of callus, which was yellow in colour and granular. This callus turned greenish granular by 18 days, Small leaflets were arising around 19 days, showing regeneration. George and Rao (1980) have observed *in vitro* regeneration of *Brassica juncea* on cotyledonary explants on MS + NAA+ BAP + Coconut milk. In most of the treatments white hairs were observed especially in the combination of IBA + Kn, where the hairs spread profusely on the callus.

References

Bhuiyan, M.S.A., M.M. Haque, M.I. Haque, R.H. Sarkar and A.S. Islam. 1992. Morphogenic responses of Peanut leaflet explants cultured *in vitro*. *Plant Tissue Culture* (Bangladesh), **2(1)**:49–53.

Eapen, S. and L. George. 1993. Plant regeneration from leaf discs of Peanut and pigeon pea, influences of benzyladenine, indole acetic acid and indole acetic acid-amino acid conjgates. *Plant Cell tissue and Organ Culture*, **35(3)**:223–227.

George, L. and P.S. Rao. 1980. *In vitro* regeneration of mustard plants (*Brassica juncea*) var. RAI-5 on cotyledon explants from non-irradiated, irradiated and mutagen-treated seeds. *Ann. Bot.*, 107–112.

Ngo, H.L. and K.H. Quesenberry. 2000. Day length and media effects on *Arachis pintoi* regeneration *in vitro*. Fifty-ninth annual meeting of the Soil and Crop Science Society of Florida, U.S.A. 22–24 September, 1999. Proceedings, **59**:90–93.

Radhakrishnan, T., T.G.K. Murthy, K. Chandran and A. Bandyopadhyay. 2000. Micropropagation in Peanut (*Arachis hypogea* L.). *Biologia Plantarum*, **43(3)**:447–450.

Sharma, K.K., S.S. Bhojwani and T.A. Thorpe. 1991. The role of cotyledonary tissues in the differentiation of shoots and roots from cotyledon explants of *Brassica juncea* (L.). *Plant cell tissue organ culture*, **24**:55–59.

Venkatachalam, P., and N. Jayabalan. 1997. *In vitro* response of different explants on callus development and plant regeneration in Groundnut (*Arachis hypogaea* L.). *Crop Improvement*, **24(2)**:167–173.

Vibha Tyagi, Parul Sharma, P.L. Swarnakar, V. Tyagi and P. Sharma. 1993. Initiation of callus and differentiation of multiple shoots *in Arachis hypogaea* L. *Ann. Bot.*, 34–37.

Mycotoxins Problems in India, their Management — A Review

*P. Giridhar** and *S.M. Reddy***

ABSTRACT

Mycotoxin producing moulds tend to spread aggressively on food stuffs and can cause mycotoxicosis at epidemic scale due to systemic and variety of toxic activities of mycotoxins in animal systems. The direct food loss can be estimated from the value of rejected mycotoxin contaminated foods and feeds. Thus poorer their market value and rebuff of such exported foods/feeds lead to severe financial losses which inturn affects the economy of the nation. An overview is presented covering the incidence of mycotoxigenic fungi and mycotoxin contamination of various foods and feeds and mycotoxicoses outbreak of animals including man in India and their management.

Introduction

Human activities affect both the size and structure of fungal populations from a human perspective. Most fungi associated with cultivated crops dictate human activity inadvertently. However, they partly dictate which and how many fungi occur and how the fungi both directly and through metabolic products influence human activities, domestic animals and then humans themselves (Cotty *et al.*, 1994).

With the advent of green revolution in India, food production had been achieved. The growth of Indian agriculture still continues to be linked with the vagaries of nature. Certain fungal pathogens which were causing minor diseases during earlier years have become severe diseases and in some cases they assumed epidemic proportions. Nearly seventy percent of the

1 Plant Cell Biotechnology Department, Central Food Technological Research Institute, Mysore-570 020, India.

2 Department of Botany and Microbiology, Kakatiya University, Warangal-506 009, India.

total production of food grains in India is retained at farm level where the poverty, unscientific and faulty storage conditions enhance the chances of mould infestation and thereby mycotoxin contamination. Storage moulds and bacteria are present in an inactive state in the field and transport on food grains in dormant conditions (usually as spores) and grow under favourable storage conditions of food grains (Hussein and Brasel 2001). During warm, humid conditions, several of the fungi such as species of aspergilli, fusaria and penicillia increase rapidly in association with grains (Girish and Goyal 1986). The activity of fungi there may cause decrease in germinability, discoloration of food grains, heating and mustiness, loss in weight, along with biochemical changes and production of toxins (Gajendra Prasad *et al.,* 1993). All these changes may occur before the responsible fungi could be detected on visual examination.

The fungi may produce a variety of mycotoxins in food grains, spices and dry fruits (Giridhar and Reddy 1997a, 1999a), vegetables, feeds and fodders and their products (Laxma Reddy and Reddy 1992) either at pre-harvest or post harvest. Because of their potent toxic nature and adverse effect on animals including man and crops (Bilgrami *et al.,* 1986; Bhat *et al.,* 1997a,b) and results in illness and economic losses are of fairly high. The diseases or physiological abnormalities resulting due to the ingestion of mycotoxins are known as "mycotoxicoses" (Bilgrami and Sinha 1984a). In India aflatoxins, ochratoxins A, trichothecenes, zearalenone, tremorgenic toxins and ergot alkaloids are the mycotoxins of great agro-economic importance.

Genesis of the Problem

Presence of mould produced toxins in food commodities has been known since Biblical times but their role in inciting disease syndromes was realized only when it was discovered that "ergotism" was caused due to the consumption of barley and rye infected with *Claviceps purpurea.* Cardiac beriberi caused by *Penicillium citriovirde* was recorded in Japan due to consumption of contaminated rice; Stachybotryotoxicosis of horses due to *Stachybotrys atra* infestation in Soviet Union in 1931; Red Mould Diseases or Black Spot Disease caused by *Fusarium* sps. in Japan during 1940–50 and Alimentary Toxic Aleukia (ATA) caused by *Fusarium* and *Cladosporium* species in USSR during 1942–47 are some of the maladies in man and livestock. However, the severity of mycotoxin problem was realized during World War II when Russians eating mouldy over wintered grains suffered with severe dermal necrosis, leukopenia, haemorrhages and destruction of bone-marrow. *Fusarium* was found to be the causal organism. World-wide scientific recognition of mycotoxin problem was, however, only in 1960 when it was discovered that the aflatoxins were responsible for the mass death of about a lakh turkey poults (Turkey × disease) in England, due to the ingestion of the "poisonous feed" which caused the havoc was due to the feed imported from Brazilian peanuts, contaminated with *Aspergillus flavus* (Sargeant *et al.*, 1961). Discovery of incidene of fumonisins produced by *Fusarium moniliforme* in maize and their association with equinencephalomalacea in 1988 (Gelderblom *et al.*, 1988) is another event which awakened world recognisation to this problem. Marasass (2001) aptly states that if the new era of research on food-borne diseases caused by mycotoxins was introduced by aflatoxins in 1960, the twentieth century closed with world wide attention focused on fumonisins. The next chapter to be written on a food-borne diseases caused by *Fusarium* mycotoxins may well be called by Fumonisins, food borne fusarium toxins for third millennium.

Mycotoxicoses Outbreak

In India first outbreak of mycotoxicoses has occurred in 1958 as a result of ingesting pearl millet containing 15–26 mg/kg ergoline, a clavine-type ergot, from *Claviceps fusiformis*. Subsequently several disease outbreaks have been reported due to mycotoxins during last 40 years.

1. The outbreak of aflatoxin hepatitis in 1974–75 in western India claiming more than 100 tribal lives due to consumption of maize containing aflatoxins (Krishnamachari *et al.*, 1975, 1977).
2. The outbreak of enteroergotism in Rajasthan, Maharashtra and Gujarat in 1976 due to consumption of bajra contaminated with ergot.
3. Occurrence of Degnala disease, a disease of an obscure etiology, with syndromes stimulating chronic ergotism in buffaloes and cattle in 1970's in the rice growing areas of the Punjab and Haryana states (Karla and Bhatia 1990).
4. Aflatoxin and Indian Childhood cirrhosis (Bhandari and Bhandari 1980)
5. Endemic familial arthritis in India due to fusarial toxins
6. "Kodua poisoning" in man due to presence of CPA in Kodo millet (*Papsalum scrobiculatum*) produced by associated fungi (Rao and Hussain, 1985).
7. Deoxynivalenol mycotoxicosis in Kashmir valley in 1987 caused by the consumption of mould — damaged wheat products (Bhat *et al.*, 1989a).
8. Inclusion body hepatitis in poultry in Northern India from 1990–1994 (Singh *et al.*, 1994).
9. A food-borne disease outbreak due to the consumption of moldy sorghum and maize containing fumonisin mycotoxins (Prathapkumar *et al.*, 1997).
10. Severe aflatoxicosis outbreak in chicken in Himachal Pradesh (Rajesh Chahota *et al.*, 2000).

Natural Incidence of Mycotoxins in different Food Commodities

Cereals constitute the most important food and feed sources, which are affected by different mycotoxigenic fungi. The problem of natural contamination of mycotoxins in cereals aggravated due to rapidly changing agricultural technology. Though in general mycotoxins seem to pose great threat in the tropics than in the temperate regions, but no part of the world can be considered to be mycotoxin-free zone due to the movement of various food stuffs from one part of the globe to the other (Bilgrami and Sinha 1984a). Some of the important commodities, which have been reported to be naturally contaminated with one or the other mycotoxins are listed in table 1.

Table. 1 Incidence of mycotoxigenic fungi and mycotoxins in different foods and feeds from India

Item	*Fungi / mycotoxin*	*Reference*
Apple	AFB1	Roy & Chourasia (1989), Neelakanthan *et al.* (1984)
Anise fruits	Trichothecenes	Ghosal *et al.* (1982)
Barley	Mycotoxins	Mall *et al*(1979)
Beetle nut	AFB1	Raisuddin and Misra (1991)
Black cumin	Mycoflora and mycotoxins	Giridhar and Reddy (1997a)
Black pepper	Mycoflora and mycotoxins	Giridhar and Reddy (1997a,e, 2002)
Bread	AFB1	Gour and Siradhana (1989)
Cereal grains	*A. flavus*, AFB1	Sashidhar *et al.* (1989)
Chick pea seeds	Mycoflora and aflatoxins	Ahmad and Singh (1991)
Chilli (dry)	Mycoflora, mycotoxins	Dandavate and Shinde (1986), Giridhar and Reddy (1996), Reddy *et al.* (2001)
Coconut copra	Mycoflora and mycotoxins	Giridhar and Reddy (1996b; 1997a)
Coriander	Mycotoxins	Misra 1987; Tirumala Devi *et al.* (2001)
Date fruit	Mycoflora, mycotoxins	Giridhar and Reddy 2001b
Dry fruits	Mycoflora and mycotoxins	Saxena and Mehrotra, 1990; Giridhar and Reddy (1999a, 2001a, b, 2002)
Fennel seeds	Mycoflora, mycotoxins	Misra and Batra (1987)
Fish	AFB1	Prasad and Sinha (1987)
Feed & fodders	Mycotoxins	Pruthi (1978), Neelakanthan *et al.* 1978; Laxma Reddy and Reddy (1992a), Reddy and Girisham (1993); Sinha *et al.* (1999)
Fenugreek	Mycoflora and mycotoxins	Giridhar and Reddy (1997d, 1999c)
Garlic and onions	Mycoflora and mycotoxins	Giridhar and Reddy (1999d)
Ginger	Mycoflora and mycotoxins	Giridhar and Reddy 1997a

Item	*Fungi / mycotoxin*	*Reference*
Groundnut cake	AFB1, T-2 toxin, other mycotoxins	Subramaniyan and Rao (1974), Rukumini and Bhat (1978), Mehan and McDonald (1984), Nagaraj and Kumar (1986), Bhavanishankar and Shantha (1987), Rati and Shanta (1994), Kaur *et al.* (1994), Bhat *et al.* (1996), Monteiro *et al.* (1996), Usha *et al.* (1994), Sashidhar (1993),
Gram flour	Mycotoxins	Sinha and Sinha 1991, Ramakrishna *et al.* 1990,
Herbal meidicines	AFB1	Kumar *et al.* (1993)
Linseed	AFB1	Sahay *et al.* 1990
Macuna puriens	AFB1	Roy and Chourasia (1989)
Makhana	Mycotoxins	Sinha *et al.* 2002
Maize	AFB1, Mycotoxins	Bilgrami and Sinha (1982, 1983); Bhavanishankar and Shantha, (1987) Sinha and Ranjan (1989), Sinha (1987; 1988), Krishna Reddy and Reddy (1990); Vasanthi and Bhat (1997), Vasanthi *et al.* (1997); Bhat *et al.* (1997a), Bilgrami *et al.* (1992), Janardhana *et al.* (1999)
	FB1 Cyclopiazonic aicd (CPA)	Chatterjee and Mukherjee (1994), Krishna Reddy and Reddy (1992)
Mango	Aflatoxins, Fumonisins	Ghosal *et al.* (1977)
Macuna	AFB1	Roy and Chourasia (1989)
Medicinal plant seeds	AFB1	Roy and Kumar (1993)
Milk and milk products	Mycotoxins	Sinha and Ranjan (1991); Manorama and Sinha (1995)
Mustard	Mycotoxins	Sahay and Prasad (1990), Ranjan *et al.* (1992), Giridhar and Reddy (1997a,b)
Oil seeds	Ochratoxin A, fusarial toxins	Rati *et al.* (1979); Chakrabarti (1987) Rati and Shantha (1994)
Paddy	Mycoflora and mycotoxins	Chenulu and Dharam vir (1979); Gnagopadhyay and Chakraborty (1981); Chary and Reddy (1987a); Waghray *et al.* (1988), Jayaram and Kalyanasundaram (1990); Begum and Samajpathi (2000); Sudhakar Rao and George (2000)
Poultry feeds and foods	Mycotoxins	Gopal (1985); Nagaraj and Kumar (1986), Balachandrana and Ramachandran (1987); Surekha and Reddy (1990), Dhavan and Choudary (1995); Balachandran and Parthasarathy (1996); Prathapkumar *et al.* (1997), Shetty and Bhat (1997a,b); Jindal *et al.* (1999), Suman Mor and Kishan Singh (2000)

contd...

...*contd*. Table 1

Item	*Fungi / mycotoxin*	*Reference*
Pearl millet	Mycotoxins	Girisham and Reddy(1988); Misra and Daradhiyar 1991, Reddy and Nusrath 1986, Sinha 1988
Pulses	Mycoflora and mycotoxins	Premalatha Singh *et al*. (1990); Begum and Samajpathi (2002)
Papasalum (Kodua millet)	CPA, mycotoxins,	Rao and Hussein (1985), Dwivedi *et al*. (1990)
Quinces (*Cydonia* fruits)	AFB1	Sharma and Sumbali (1999)
Raisins	Mycoflora and mycotoxins	Giridhar and Reddy 2001a
Safflower seeds	Trichothecenes	Ghosal *et al*. 1977
Scented supari	Patulin	Neelakanthan *et al*. (1982)
Sorghum	Mycoflora, mycotoxins	Tripathi (1973); Rukmini and Bhat (1978); Reddy and Nusrath (1985); Sashidhar e t al. (1992a); Mukherjee and Lakshminarasimham (1995); Hareeshvardhan Rao *et al*. (1996a, b, c, 2001a, b); Shetty and Bhat (1997a)
Stored grains	Mycotoxins	Singh *et al*. (1982, 1983)
Sunflower seeds	Mycoflora and mycotoxins, Toxigenic fungi	Vijaya Lakshmi and Rao (1985).
Turmeric corms	Mycoflora and mycotoxins	Giridhar and Reddy 1997a
Tobacco (chewing type)	AFB1	Suman Rai and Chaturvedi (1993)
Wheat	Mycoflora and mycotoxins	Augustine *et al*. (1984)
Air borne aflatoxin in grain processing industries	AFB1 incidence	Ghosh *et al*. (1991)

Maize

Maize is an excellent substratum for mould growth and mycotoxin production. In India, systematic survey of maize grains for mycotoxins contamination has been undertaken in some parts only. Some reports mainly focused on aflatoxin contamination in stored grains have indicated that the problem of aflatoxin contamination is much more serious than usually visualized (Bilgrami and Sinha 1984a, b). In an extensive survey (1977–1980) of the important maize growing areas of Bihar State, natural contamination of mycotoxins and particularly of aflatoxins was reported in the grains of maize crop still in the field (Bilgrami *et al*., 1981a, b, c). Apart from this out of 615 samples of maize 316 samples contained aflatoxin B1 (AFB_1) in concentration ranging from 8 to 1640 ppb. Surveys conducted by the Indian Grain Storage Institute during 1978–79 and

1980–81 indicated contamination. Rati *et al.* (1979) reported the presence aflatoxins in the range of 30–50 ppm of ochratoxin A in corn samples collected from Hapur, U.P. and Mysore, Karnataka State respectively.

Under a FAO sponsored Food Contamination Monitoring Project in the year 1980, out of 10 maize samples collected from Western Uttar Pradesh, 3 samples were contaminated with AFB_1 in the range of 20–80 ppb (Indian Grain Storage Institute, Hapur). Similarly samples of maize collected from traditional storage structures from different parts of India have shown contamination of AFB_1 up to 15,680 ppb (Singh *et al.*, 1982). Aflatoxin production by *Aspergillus parasiticus* in cereals leads to loss in caloric value of maize seeds (Premlatha Singh 1988). A positive correlation of insect damage and aflatoxin contamination of maize crop was observed by Sinha and Ranjan (1989). The incidence of aflatoxins, zearalenone (Bilgrami *et al.*, 1980a), trichothecenes (Krishna Reddy and Reddy 1989), cyclopiazonic acid (Krishna Reddy and Reddy, 1992) on standing crop of maize was also reported. Sinha (1990) reported the heavy infestation of maize samples of Kharif crop, which contained more aflatoxins (47%) than the samples of the rabi crop (17%). Bhat *et al.* (1997) reported out of 2074 samples of maize from rural and urban areas of 11 states of the country 26% of maize samples contained a aflatoxins beyond the permissible limit (30 mg/kg). Analysis of aflatoxin intake from maize based diets in a rural population in southern India by Vasanthi and Bhat (1997) revealed that the average afaltoxin intake in house ranged from 0.33 to 1.5 mg/day, while individual intake varied from 0.08 to 2.22 mg/day. They further recorded a considerable variation in fumonisin B1 (FB_1) contamination between the hybrids and inbreds.

Rice

Natural occurrence of mycotoxins especially aflatoxins and aflatoxin producing fungi in rice has been reported from different parts of the world (Bhat *et al.*, 1978). They identified coastal areas of A.P. and cauvery delta of Tamilnadu as areas, which are prone to cyclonic weather conditions during paddy harvesting season as high risk areas of mycotoxin contamination and reported aflatoxins, zearalenone and sterigmatocystine in different paddy samples. Joshi *et al.* (1987a) reported that AFB1 in the range of 40 ppb in paddy harvested from rain affected crop in Punjab and Haryana. Aflatoxin contamination of rice has also been reported from different parts of India (Tyagi *et al.*, 1986). Bilgrami and Sinha (1984b) is of the opinion that though presence of mycotoxins like citrinin, sterigmatocystin and ochratoxin A has been detected as natural contaminants of rice in one or the other region of the world, only scanty information is available pertaining to these toxins in India. Dehusked rice collected from the flood affected areas of Warangal district of A.P. (Chary and Reddy 1987a and b) and other parts of A.P. (Nusrath and Ravi 1984) were contaminated with different mycotoxins.

Wheat

Wheat which contributes 36% to our country's total food grain basket, does not appear to be a good substratum for aflatoxin elaboration. Srivastava (1987) reported that out of 223 samples collected from wheat growing belt of Western U.P., only 9 samples were contaminated with aflatoxin B1 in the range of 8–40 ppb (Vasan, 1980; Bilgrami and Sinha, 1984a). All the six flood affected wheat samples collected from Punjab contained aflatoxin B1 in the range of 8–40 ppb (Joshi *et al.*, 1987b). Vora (1978) reported the natural incidence of aflatoxin and sterigmatocystine respectively. Rati *et al.* (1979) reported the presence of 30–50 ppm of ochratoxin A in wheat samples collected from Mysore. The incidence of enterogastric disorder outbreak in Kashmir valley during July–September 1987 was attributed to trichothecenes contaminated wheat products (Bhat *et al.*, 1989a, b). Most of the wheat samples contained deoxynivalenol in the range of 0.43–4.85 μg/g, followed by acetyl deoxynivalenol (0.6–2.5 mg/g), nivalenol 90.1 μg/g) and T2 (4 μg/g) .

Barley, oats and other micro millets

Micro millets also do not favour natural contamination of mycotoxins. However, bajra has been reported to be suitable substratum for aflatoxin production. Presence of cyclopiazonic acid (CPA) in Kodo millet (*Papsalum scrobiculatum*) causing "Kodua poisoning" in man was reported by Rao and Hussain (1985).

Sorghum

Sorghum is prone to head moulds infestation and mycotoxin contamination. Bhat (1988), Ramakrishna *et al.* (1990) recorded to the extent of 24.5 mg/kg of T2. Rukmini and Bhat (1978) recorded T2 toxin in *Fusarium incarnatum* infested sorghum. Similarly the production of DON by *Fusarium* isolates of wheat and sorghum was reported to be associated with human mycotoxicosis outbreak (Ramakrishna *et al.* (1989). Reddy and Nusrath (1985) reported the infestation of different varieties of sorghum seeds collected from A.P. by mycotoxigenic fungi and mycotoxin contamination. Rati *et al.* (1979) reported the presence of 50–70 ppm of ochratoxin A in sorghum samples collected from Mysore. In Western India Sorghum samples were contaminated with AFB_1, depending on seed moisture (Mall *et al.*, 1986). A extensive study under taken by NIN, Hyderabad in A.P.

revealed that 26 isolates of sorghum belonging to different species of *Fusarium* such as *F. proliferatum* and *F.subglutinans* produced fumonisin B1 up to 2 µg/g which is significantly low compared to maize (93–218 ug/g). The natural incidence of *Alternaria* mycotoxins i.e. alternariol (AOH), alternariol methyl ether (AME), alternuene (ALT), altertoxin-1 (ATX-1) and tenuazonic acid (TA) in sorghum has also been recorded by Ansari and Srivastava (1990).

Pulses

Pulses being proteinaceous in nature constitute a good substratum for mycotoxin producing fungi. However, there are only a few reports about the natural occurrence of mycotoxins in these substrates. Association of toxigenic fungi with some samples of pulses has been confirmed by Vora (1978). A survey under FAO sponsored by Food Contamination Monitoring Project in Western Uttar Pradesh revealed that few samples of pulses including green gram, black gram, lentil and few samples of cotton seed were contaminated with AFB_1 in the range of 4–80 ppb and 35–200 ppb respectively. Similarly, ten percent of stored soybean samples from Uttar Pradesh were contaminated with aflatoxin B_1 in the range of 4–40 ppb (Joshi *et al.*, 1986b). The incidence of mycotoxin producing fungi in black gram, red gram and green gram collected from different regions of Andhra Pradesh State (A.P.) was dominated by species of *Aspergillus* and *Penicillium* sp.

Premalatha Singh *et al.* (1990) reported nutritional deterioration and presence of aflatoxins in some pulse cultivars under the influence of *Aspergillus flavus*. Premalatha Singh and Ahmad (1989) recorded aflatoxin in seed samples collected during the development of pod in the range of 30–110 µg/kg.

Oil Seeds

Groundnut

Ground nut is one of the most vulnerable commodity for *A. flavus* infestation and AFB1 contamination. Rati *et al.* (1979) recorded 50–2000 ppb of ochratoxin A in ground nut samples collected from Mysore region. Bhavanishankar and Shantha (1987) reported the natural incidence of *Fusarium* toxins in peanut samples collected from Mysore. Of the total number of peanut samples analyzed, 6.9 % were positive for the toxic trichothecenes (T-2 up to 38.89 mg kg^{-1}; DAS up to 2.03 mg kg^{-1}). Relative humidity during storage is reported to play pivotal role in mycotoxin contamination of ground nut (Gambhir and Khairnar, 1986). A multicentric study carried out by the FAO under the Food Contamination monitoring

Project (FCMP) in India indicated that groundnut is most prone to aflatoxin contamination. Various surveys carried out in different parts of India also point to ground nut as a high-risk commodity for aflatoxins (Bhat, 1989a). The levels of aflatoxin in unrefined oil in India varied, the highest recorded level being 5000 μg kg^{-1}. Assuming a realistic average toxin level of 80 μg kg^{-1}, use of 50 g of unrefined groundnut oil per day has been calculated to contribute 4 μg of aflatoxin to the diet per day (Acharya, 1986). A multicentric study for aflatoxin B1 contamination in ground nut samples from different geographical regions of India revealed the presence of aflatoxins B1 in the range of less than 5 to 833 mg/kg (Bhat *et al.*, 1996). The Codex Alimentarius Commission Working Group recommended aflatoxin limit of 5 mg/kg in nuts was exceeded in 44.9 % of samples.The trichothencenes and zearalenone were reported to contaminate groundnut oil cakes (Rajakumari Kaushal *et al.*, 1994) and sesamum (Reddy and Reddy, 1983). The levels of aflatoxins were about 31% in sweet and spiced peanut products collected from shops (Rati and Shantha 1994). However, aflatoxin levels in the toffees were lower (2–400 ppb). In 1994–95 a large quantity of ground nut was rejected as unfit for export due to the presence of aflatoxins. Rejection of groundnut for export was attributed to aflatoxin B1 contamination and inappropriate sampling techniques.

Mustard

Mustard is also susceptible to mycotoxigenic fungal infestation (Giridhar and Reddy 1997a). Survey of Incidence of mycotoxin producing fungi on mustard in Bihar by Sinha *et al.* (1988) and Sahay and Prasad (1990) and Bilgrami *et al.* (1991) indicated the aflatoxin contamination in the range of 750, 87, and 1420 μg/kg respectively in 80% of samples. Mycotoxins producing fungi can damage mustard seedlings (Giridhar and Reddy, 1996, 1997a). The incidence of aflatoxin in mustard seeds, cakes and oil was also reported from other parts of India (Sahay, 1988).

Dry fruits and spices

Aflatoxin has been reported as the main contaminant of coconut and other dry fruits viz., areca nut, almond, cashew nut, walnut, raisin, makhana and emblic (Chaturvedi and Chaturvedi, 1992; Giridhar and Reddy, 1999a,b). In some samples of these dry fruits, presence of zearalenone, citrinin and ochratoxin A has also been reported (Giridhar and Reddy, 1996a, 1999a, 2001a, b). Natural contamination of aflatoxin has been reported in different samples of spices by Madhyasth and Bhat (1984), chilli (Jyothi Saxena *et al.*,1988; Giridhar and Reddy, 1999b), ginger (Giridhar and Reddy 1996), cardamom and turmeric (Giridhar and Reddy, 1997a) has been reported from different parts of India. Presence of ochratoxin A, citrinin and zearalenone has also been reported in a few samples of spices (Anjana Singh 1983; Bilgrami 1984, 1985). In a survey in Uttar Pradesh conducted by the Indian Grain Storage Institute, Hapur in 1984, out of 99 dry fruit samples only 19 samples contaminated with aflatoxin B1 in the range of trace to 50 ppb. Production of aflatoxins in mustard by isolates of *Aspergillus flavus* was reported by Anisa Ather *et al.* (1988) . Aflatoxin was one of the contaminants of coconut copra (Krishna kumari *et al.*, 1984 and Giridhar and Reddy 1997a) of A.P. and Maharashtra State (Kulkarni *et al.*, 1986) and other parts of country (Narasimhan, 1968). Natural incidence of mycotoxigenic fungi and mycotoxins has been reported on bitter gourd seeds (Rajendra Singh *et al.*,1991). Survey for the incidence of mycotoxigenic fungi in different spices indicated that 48.07% of black pepper, 31% of caraway, 30% of dry chilli, 28% of methi, 60% of ginger, 35% of turmeric samples contained one or other mycotoxins produced naturally. These were found to contain aflatoxins, sterigmatocystin, patulin, citrinin, cyclopiazonic acid were encountered (Giridhar and Reddy, 1997a). Natural incidence of mycotoxins in date fruits, areca nut, raisins, coconut copra, cardamom, cashewnut etc were good substrates for mycotoxingenic fungi and toxin production under faulty storage conditions (Giridhar and Reddy 1996a, b, 1997a, 1999a, 2001a, b). Variation in the incidence of mycotoxigenic fungi in different agricultural commodities may be attributed to suitability of substratum nature, climatic conditions, storage practices, temperature and high humidity (Giridhar and Reddy, 2002).

Milk and Milk Products

Available evidence indicates that aflatoxin B1 present in groundnut meal, one of the major ingredients of cattle feeds, not only causes liver damage in milch cattle but could also be excreted in milk as such or in the form of aflatoxin M1 (Lewis *et al.*, 1994). Yadagiri and Tulpule (1974) reported aflatoxin M, up to 8 mg/l in the buffaloes milk. Outbreaks of aflatoxicosis in dairy cattle from Mysore region and in Murrah buffaloes from Andhra Pradesh region emphasize the need for a constant awareness and monitoring.

Feeds

Mycotoxins in the feed are exceptionally important in animals reared under intensive husbandry conditions. Aflatoxin in feeds is still the prima concern because of wide range of host susceptibility, ubiquitous presence of aflatoxigenic fungi and conducive growth conditions, immunosuppressive properties, and heat stable properties of aflatoxins (Lillehoj 1973). Usually, the food which is declared unfit for human consumption finds its way as feed for animals and poultry birds (Bilgrami and Sinha, 1984b). Incidence of mycotoxicosis in poultry due to the ingestion of contaminated feed has been reported from different parts of India (Neelakanthan *et al.*, 1978; Sivaswamy *et al.*, 1991; Singh *et al.*, 1994; Dhavan and Choudary 1995; Balachandran and Parthasarathy 1996; Jindal *et al.*, 1999). Still many times they go unnoticed, as proper reporting is not being done. Similarly the incidence of toxigenic fungi associated with poultry feed have been reported very frequently (Neelakanthan *et al.*, 1978; Surekha and Reddy, 1990, 1991; Laxma Reddy and Reddy, 1992). Survey on the presence of fusarial toxins, sterigmatocystin etc in animal feeds in Tamil Nadu State of India was reported by Department of Food Technology, Tamil Nadu Agricultural University in 1980. Out of 52 strains of *Fusarium* screened 36 were positive for one or other trichothecene group of fungi on cattle feed samples collected from different regions of Andhra Pradesh (Surekha and Reddy 1990).

Fodders

Straw of rice, maize and sorghum left after harvesting of the grains are the main bulk of fodders for livestock throughout India. Tropical conditions, harvesting practices, unseasonal rains, high moisture content of straw during baling and open storage not only promote fungal proliferation but also encourage mould growth and to elaborate many mycotoxins.

Sea foods

Gayatri Murjani (2000) recorded remarkable gross changes in various internal organs of fishes (Indian major carp *Labeo rohita*) that received AFB1. Verma and Pandey (1987) and Verma *et al.*(1988) have studied the effect of aflatoxin B1 on the liver of an air breathing teleost, *Channa punctatus*. Studies proved that aquatic animals are sensitive to carcinogenic stimuli and can serve as indicators of carcinogenic hazards to man (Shank *et al.*, 1972). It has also been shown that livestock products like meat, milk and eggs do contain aflatoxin in the form of residue (Rajan, 1995).

Apart from the above discussed reports on the incidence of myctoxins on agricultural produce, some other reports on other foods were given in Table 1.

High Risk Areas for Mycotoxins in India

Climatic conditions in India are most conducive for mould invasion, proliferation and elaboration of mycotoxins. Moisture content of the grain or the relative humidity surrounding the substrate are the most important factors governing the aflatoxin production by *A. flavus*. The high risk areas identified in India are Kerala, Western India, Gangetic plains, north eastern as well as coastal areas of Andhra Pradesh, Karnataka and Tamil Nadu (Bilgrami 1984). Unseasonal rains and flash floods which are very common in India enhance the moisture content of the grain making them more vulnerable for fungal attack (Srivastava 1987).

Weather Conditions and Mycotoxins

Warm humid climate provides congenial atmosphere for the growth of fungi and production of mycotoxins. Food grains are normally harvested at higher moisture content and then dried to bring down the moisture content up to safe level (less than 10%) before storage. Delay in drying to safe moisture levels increases risks of mould growth and mycotoxin production (Srivastava 1987). Natural calamities like floods or torrential unseasonable rains during pre, mid or post-harvest stages may render the crops vulnerable to microbial attack.

Faulty storage conditions may also enhance the chances of microbial attack and production of mycotoxins. Thus, starting from harvesting of the crop till the food or food products are consumed by the consumers, fungus may invade at any stage/step. Two important environmental factors that, govern mould infestation are substrate water activity and temperature. The minimum a_w at which molds are capable of growing is approximately 0.61. Further the a_w and temperature limits for growth and mycotoxin production are some times markedly different. Giridhar and Reddy (1997a) have discussed elegantly the role of water activity and temperature in production of mycotoxins on mustard, maize etc. Apart from this, the insects that attack the crops at pre and post-harvest conditions are responsible for *A. flavus* infestation and aflatoxin production (Sinha and Sinha, 1992).

The increased contamination of maize with fumonisins, AFB_1, and wheat samples with AFB_1, DON, Zearalenone etc in India may also attributed to drought conditions. Furthermore, the demand for increased food grain production has paved the way for changing farming pattern may also contribute to the severity of mycotoxin problem.

Management of Mycotoxins

Mycotoxins are one of the important factors contributing for food loss. Based on the Global Environment Monitoring system — Food Contamination Monitoring and Assessment Programme (GEMS/FOOD) data, mycotoxin contamination is a widespread problem of the food supplies in most countries. As a result, many countries have enacted regulations to control level of mycotoxins in the national food supply as well as in food moving in international trade. Eighteen countries all over the world have guidelines or regulations, which prescribe maximum acceptable limits for mycotoxins in foods and feeds. The limits prescribed for aflatoxins vary from 0 to 50 ppb in foods and from 0 to 1000 ppb in feeds in different parts of the world (Srivastava 1987). In India the permitted levels of AFB_1 in food commodities is 30 ppb but no such regulations are prescribed for the other mycotoxins. Various analytical methods for the screening and analysis of mycotoxins are given in Table 2. It is time to realize the problem and is enact wisely by making regulations for these mycotoxins also to avoid health risks of man. In general all the agricultural commodities before export are subjected to stringent analysis of mycotoxins such as aflatoxins. Though health hazards from aflatoxins in developing countries such as India is more from maize than groundnut, the economic significance of aflatoxin contamination in terms of reduced export earnings and loss at different levels of the trade chain such as shelled groundnuts, edible oils, and oil cakes is more in groundnuts than in maize (Bhat, 1989b). Hence, the Governmental agencies should procure food grains confirming to prescribed specifications thereby minimizing the chances of contamination and thus ensure the supply of quality food grains to the consumers. At farm level also quality consciousness should be created amongst the farmers through a network of expert teams of Save Grain Campaign.

There is an urgent need to form a National Guidelines for prevention and control of mycotoxins contamination specially aflatoxin in agricultural commodities, which should include: harvesting of fully matured grains/cobs/ears and drying to safe moisture level (less than 10%), separating the healthy, sound grains/cobs/ears from damaged or rotted or infected,

Table 2: Analytical methods developed for mycotoxins analysis from India

Area of research	*Reference*
Modified Pon's method for estimation of AFB1.	Rati *et al.* (1987)
Single test tube screening method for AFB1 detection	Pallavi *et al.* (1997)
Screening method for detecting ergosterol in cereals	Sashidhar *et al.* (1989, 1991)
A method for detection of DON	Ramakrishna *et al.* (1989)
Evaluation kit for AFB1 at field level	Sudershan *et al* (1992)
p-tolualdehyde test for penicillic acid detection	Neelakanthan *et al.* (1978)
Chemical changes in maize seeds & apples infected with *A. flavus*	Bilgrami *et al* (1981c); Sinha and Singh (1987)
Colorimetric method for citrinin	Damodaran *et al* (1973)
Colorimetric method for terreic acid	Subramanian *et al* (1978)
Colorimetric method for patulin	Subramaniyan (1982)
Spray reagent for trichothecenes	Rao *et al* (1985)
Colorimetric method for CPA	Rathinavelu and Shanmugasundaram, (1984)
Minicolumn for the duration of sterigmatocystin	Ramakrishna and Bhat (1990)
Detection of aflatoxins by pressure minicolumn technique	Sashidhar *et al.* (1988)

drying the grains/cobs/ears on raised bamboo floors along with other appropriate drying methods as required for the merchants (for ex. Dry shelled maize on concrete drying floors), using a water proof sheet (plastic not advisable) to cover drying produce for rain protection, even mechanical drying if required, storage at proper dry conditions, along with good ventilation, efficient sanitation practices to destroy or discard infested ears/cobs/ etc., and effective fumigation for control of insect problem.

Prevention of the growth of moulds, separation of infected grains, detoxification and growing resistant varieties are some of the other strategies for mycotoxin contamination (Bilgrami and Sinha 1984b).

Prevention of mould growth

Mould damage in stored grain, may be prevented mainly by three kinds of treatments viz., drying of grain, controlled storage atmosphere and chemical treatment.

Drying of grain

It is an established fact that dry grain stores long, safe from insects and moulds because the requirements of moisture for their growth and development are not met. Exposure of aflatoxin contaminated groundnut oil to sun light has given very promising results as it destroyed about 99 percent of the aflatoxins (Bilgrami and Sinha, 1984b). Other studies suggested that urea and sunlight can detoxify aflatoxin in groundnut flour and cake (Shantha *et al.*, 1986). Other methods of grain drying include mechanical drying, in-bin drying, infrared, microwave or sonic and solar energy drying (Srivastava, 1986).

Controlled atmosphere storage

For most of the fungi especially aflatoxin producing fungi requires warm humid conditions (32°C is the optimum or 30–40°C is the range, RH of 85% or 62–99% range). While for Fusarial toxins the *optimum* temperature is 15–28°C. So proper storage conditions are required. The significance of underground storage may prove to be safe which results in grain cooling and depleting the oxygen content to the desired level whereby the microbes and insects cannot grow.

Chemical treatment

Chemical control of fungal deterioration of stored grain is restricted to the treatment of grain for seed purposes only but not for food and feed. Kubena *et al.* (1993) reported the degradation of aflatoxin and diacetoxy nivalenol with hydroxy Na-Ca alumina silicate. Experiments conducted by Indian Grain Storage Institute showed that "Grain treat" (Mixture of propionic acid, acetic acid and benzoic acid) against species of *Aspergillus* and *Penicillium* failed to give effective protection. Antioxidants like butylated hydroxy anisole (BHA), propyl gallate (PG) could check the production of citrinin by *Penicillium citrinum* (Giridhar and Reddy 2001c). Giridhar *et al.* (1997), Krishna Reddy *et al.* (1997) reported safe management of different mycotoxigenic moulds by some of the polyamine biosynthesis inhibitors like BCHA, MGBG and DFMO. Phenol and 2,4-D inhibited the cyclopiazonic acid production by *Penicillium griseofulvum* (Krishna Reddy and Reddy 1989). Similarly Laxma Reddy and Reddy (1992) reported that p-nitrophenol and alpha-naphthol are effective inhibitors of satratoxin production by *Stachybotrys atra.* Casadio and Niola (1993) reported the destabilization of patulin in fruit juices with microwaves. Clavero *et al.* (1993) could destroy aflatoxin with the help of hydrogen per oxide. Seed treatment with Bavistin and TMTD (Trimethyl thiuram disulphide) at 0.25 per cent concentration gave 100 per cent protection to wheat grain during one year storage against species of *Aspergillus* and *Penicillium* (Srivastava, 1986). Certain mild phenolics like ferulic acid and O-Vanillin have also been reported to prevent aflatoxin production on rice, wheat, maize, groundnut and mustard seeds (Bilgrami and Sinha 1984a). The use of oxalic acid (3%) application on ground nut kernels was effective in inhibiting AFB1 production. Volatile compounds which have no residues and highly active at low concentrations are being advocated in the protection of food grains at a post harvest stage. Vapours of cyclohexane and ethyl acetate inhibited patlulin production by *Aspergillus terreus* (Girisham and Reddy 1989a), while n-butanol and formic acid could check cyclopiazonic acid production by *Penicillium griseofulvum* (Krishna Reddy and Reddy 1989). *Penecillium crustosum* and *Stachybotrys atra* were sensitive to vapours of propionic acid and failed to produce penitrem B and satratoxin (Laxma Reddy and Reddy 1992) respectively. By decreasing level of oxygen and in its place of CO_2 or N_2 are increased, creating anaerobic conditions discourages mould growth and mycotoxins contamination. Modified atmospheric packing (MAP) and controlled atmosphere storage (CAS) are to keep away mycotoxigenic moulds from stored food grains for 6–9 months. Methods adopted to control mycotoxins by different workers are précised in table 3.

Detoxification

Cooking at atmospheric pressure can destroy about 50 percent of mycotoxins. Dry roasting and oil roasting of groundnut reduces aflatoxins to a significant degree. Cooking rice under 15 Ibs. pressure for 5 minutes gave

Table 3: Management practices for mycotoxins in India

Item	*Area of research*	*Reference*
Cooking methods	Detoxification of AFB1	Fasiha Rehna and Basappa (1990), Basappa and Shantha (1996), Mitra and Joseph (2000)
Aqueous extracts of tea and coffee	Impact on AFB1 toxicity	Shantha *et al.* (1996)
Modified buchner funnel filtration	Mycotoxin decontamination process for oils	Kavita Waghrey and Uma Reddy (1996)
Impact of light	AFB_1 stability in grains	Shrivastava *et al.* (1991)
Influence of cropping pattern	AFB_1 contamination	Bilgrami *et al.* (1992)
Biological control	AFB_1 Citrinin	Shantha and Rati (1990); Choudhary (1992) Giridhar and Reddy (1997g)
Plant extracts	AFB_1	Bilgrami *et al.* (1979;1980b); Ghewande and Nagaraj (1987); Kumar and Prasad, (1992); Sinha *et al.* (1993); Prasad *et al.* (1994)
	Penitrem	Surekha and Reddy (1992)
	Satratoxin	Laxma Reddya and Reddy (1992b)
	Citrinin	Girdhar and Reddy (1996)
Food preservatives	AFB1, other mycotoxins	Chourasia (1993); Giridhar and Reddy (1996)
Cloves	AFB1 inhibition	Sinha *et al.* (1993)
storage structures on	Grains and oil seeds Black pepper	Ranjan *et al.* (1992), Giridhar and Reddy (2002)
toxic fungi	Sorghum	Hareesh Vardhan Rao *et al.* (1996c)
AFB1 production	Effect of vitamins and alcohol ferulic acid	Basappa *et al.* (1967) Bilgrami *et al.* (1981d)
	O-vanillin	Bilgrami *et al.* (1982)
	Polyamine biosynthesis inhibitors	Giridhar *et al.* (1999)
citrinin production	Polyamine biosynthesis inhibitors	Giridhar *et al.* (1997); Krishna Reddy and Reddy (1997);
	Growth regulators influence	Giridhar and Reddy (1997c)
	Vitamins and chemical mutagens	Giridhar and Reddy (1997f)
	Influence of bacteria	Giridhar and Reddy (1997g)
Terreic acid and patulin	Polyamine biosynthesis inhibitors	Giridhar and Reddy (1998)

contd...

...*contd.* Table 3

Item	*Area of research*	*Reference*
various toxigenic growth Inhibition and toxin production	Food preservatives (against CPA)	Laxma Reddy and Reddy (1992b)
	Food preservatives (against satratoxin)	Giridhar and Reddy (1996c)
	Black pepper seed	Giridhar and Reddy (2001c)
	Phenolic antioxidants	Reddy and Reddy (1990)
Interaction between various fungi	Mycotoxins production	Choudhary and Sinha (1993)
Detoxification	AFB1	Shantha and Sreenivasamurhty (1981); Sharma *et al.* (1988, 1989,)
Detoxification Biological activity	AFB_1	Bilgrami *et al.*(1984a)
	Patulin on chicks	Devaraj *et al.* (1983)
	Penitrems on chicks	Surekha and Reddy (1990)
	AFB_1 and Och A on albino rats	Rati *et al.* (1991)
	Ochratoxin A on mice	
	Aflatoxicosis in ducks	Dhuley (1997) Mukit and Kwatra (1989); Mukit *et al.* (1994)
	Laying hens fed with fumonisins	Shetty *et al.* (1997)
	Gliotoxin on chicks	Reddy *et al.* (1997)
	Vitamin A and selenium	Prakash *et al.* (2000)
	on broiler chicks fed ochratoxin A	Giridhar and Reddy (2000)
	CPA on chicks	
Influence of AFB1, CPA and satratoxin on plants	Elicitation of phytoalexins	Giridhar *et al.* (2001)
volatile compounds and environmental factors	Patulin and terreic acid	Girisham and Reddy (1989a, b)
H_2O_2 and irradiation	AFB1 inactivation	Sreeinvasamurthy *et al.* (1967); Patel *et al.* (1989)
Spice oils	Mycotoxin production inhibition	Patkar *et al.* (1994)
Pressure parboiling	AFB1, Mycotoxins	Rajendran *et al.* (1987)
Sodium bicarbonate	AFB1 inhibition	Singh and Chand (1993)

maximum destruction of aflatoxins (72 percent) as compared to ordinary cooking or cooking with excess water. Detoxification by various cooking methods is highly effective in detoxification of AFB1 (Fasiha Rehana and Basappa, 1990) . Light has also been employed successfully to destroy aflatoxin in crude groundnut oil. Studies have shown that visible light is more effective than either ultra-violet or infra-red light (Pruthi, 1978) . Similarly oven drying and autoclaving detoxified AFB1 by more than 65% (Bilgrami *et al.*, 1984b). Detoxification also could be possible by bulb and tube light treatments to the extent of 51–55% in black cumin seeds and almond respectively (Bilgrami *et al.*, 1984a). It was observed that on cooking 17–56% reduction in AFB1 concentration in maize based rotis (NIN, 1992). However, the above said drying treatment efficiency varied with mycotoxins. Some of the major cooked products such as roti and gutka (Poridge) prepared from contaminated maize and sorghum flour respectively showed about 29% reduction in extractable fumonisins B1 in case of maize and no reduction in sorghum rotis. NaCl (30%) is reported to remove FB1 from the contaminated maize (86% removal). Even the buoyant fraction of water itself removes up to 74% of FB1 when contaminated samples of maize soaked in water (NIN 1998).

Separation of infected grains

Physical separation of infested grains is an efficient and feasible method of minimizing mycotoxin contamination. This is effected either by manual operation or with the help of an electronic sorter. Fungal infection of seeds or grains usually imparts characteristic colour or other physical properties.

Resistant varieties

In recent times, efforts are being made to develop mould resistant varieties which will be mould free not only in fields on standing crop but also during storage and restrict development of moulds and mycotoxin contamination. Out of 15 commonly cultivated maize varieties screened for aflatoxin production, 4-varieties viz, Him-123, Ganga-2, Ganga 5 and EH -2420 proved to be resistant to aflatoxin production (Bilgrami *et al.*, 1980a). Similarly Krishna Reddy and Reddy (1992) during screening of maize cultivars for prevention of cyclopiazonic acid expressed its feasibility. Girisham and Reddy (1987) have reported the resistance of some of the cultivars of bajra to patulin contamination and *Aspergilllus terreus* infestation. However, evaluation and selection for resistance to mycotoxins at post harvest stage may be more difficult than it has been for plant resistance to diseases.

Integrated approach to contain mycotoxins in food and feeds is desirable. It is now realized that this approach should include efforts both to prevent contamination and to detoxify contaminated grain, especially where food resources are limited. Analytical quality assurance in agricultural export produce is required (Sashidhar *et al.*, 1992b). Biological detoxification methods such as fermentation can eliminate some classes of mycotoxins.

Biological control

Significant research has been done on biological control of mycotoxins. Fungal degradation of aflatoxins B_1 was reported by Shantha (1999). Cell free extracts of *Phoma* sp. were able to degrade AFB_1 up to 99% (50 ug/100 ml of culture media). Similarly Singh and Singh (1998) reported enzymatic degradation of aflatoxins in shake cultures of *A.flavus* CMI 102566 and highlighted the possible role of perxoidase in biological control mechanism. Karunarathre *et al.* (1990) reported that *Lactobacillus acidophilous* could prevent the growth of the moulds which in turn may pave the way for mycotoxin non production. *Marganella morganil* an intestinal organism can degrade aflatoxin (Hoh and Aibana 1982). Tuason (1993) reported that *Flavobacterium aurantiaccum* metabolizes aflatoxin. *Trichoderma viride* and *Aspergillus niger* could completely check the production of cyclopiazonic acid by *Penicillium griseofulvum* (Krishna Reddy and Reddy 1989). Girisham and Reddy (1986) have also reported effectiveness of *A. niger* in suppression of growth of *A. terreus*. Various methods for the management of Mycotoxins was given in Table 3.

Current research scenario

Development of validated new PCR assays for trichothecene and fumonisins producing *Fusarium* sps is the current trend in mycotoxins research. Though earlier investigations in India were mainly focused on incidence of mycotoxigenic fungi on various food commodities and management of mycotoxins problem, it is now finding of suitable molecular markers for mycotoxigenic fungi is the need of the day. Similarly the regulation of mycotoxin production at cellular level need to be investigated with special reference to AFB1. Jayashree and Subramanyam (1999) reported regulation of aflatoxins production by Ca^{2+}/calmodulin-dependent protein phosphorylation and dephosphorylation. The recent discussions on the *Fusarium* mycotoxins in Food Chain Project and Ochratoxin A (OTA) Prevention project of The European Mycotoxin Prevention Cluster, held in June 2003 at Uppsala, Sweden warrants the necessity of looking the mycotoxins problem in a different way with a main focus on expanding the data base on *Fusarium* toxins in cereals and aflatoxins in ground nuts, identifying critical control points in the production of maize based products and ground nut. Using natural antagonists as pre-harvest biocontrol agents against head blight and other *Fusarium* diseases. Similarly a membrane based molecular imprinted polymer (MIP) biosensors, ELISA assay and the competitive lateral flow device (LFD) are the promising tools for analysis of grain extracts for mycotoxins contamination in grain extracts, as there is a lot of significance about these mycotoxins contamination of various food commodities.

Future Prospects

As the occurrence of molds and mycotoxins in foods may lead to far reaching consequences on the health and productivity of animals including man and trade of both domestic and international. Mycotoxins research in India may be focused on the following aspects:

i) To provide effective quality control methods for domestic and export food stuffs
ii) To establish the linkage between mycotoxin exposure and human disease
iii) To establish a data base on the nature and extent of mycotoxins contamination in different foods at different stages and on the ecology of biodeterioration
iv) To promote improved post harvest management techniques for prevention of mycotoxins contamination and to improve field to storage transitions. Development of scientific and economically feasible modern scientific storage, improvement and effective marketing practices.

The contamination of agricultural produce is a pervasive and seems to be endemic problem throughout the growing regions of developing nations specially in India. It could be possible to achieve incremental reductions in crop contamination through broad modifications to agronomic practices. However, it may require a more sophisticated approach including breeding of resistant varietites and biological control methods for stable amelioration of these problems. As mycotoxin contamination occurs both at pre- and post-harvest, post- harvest reduction in food quality is of great concern. To prevent this: (i) widespread cultivation of superior locally adapted varieties, (ii) to promote adoption of appropriate crop stress management strategies, (iii) to improve post harvest management techniques and lastly to assess the potential for biological control by niche occupancy. Some of the new frontiers for mycotoxins research:

i) Application of molecular techniques,
ii) Development of portable rapid detection methods.
iii) Detection and disease phenotyping, and

iv) The discovery and characterization of biomarkers of mycotoxins are some of the important toxicogenomics and nannotechnologies may be very helpful in the management of mycotoxin problem. Molecular analysis of disease resistance in important crop plants, identification of molecular markers linked to a disease, fine mapping and cloning of genes of economic importance, development of genomic tools for the isolation of genes of agronomic interest in crops like wheat, maize etc may play a pivotal role in future in raising mycotoxin resistant varieties.

v) For the sake of farming community, consumers and exporters an authentic document is of urgent need, which should clearly meet the four important features such as (1) providing the basic knowledge regarding the mycotoxins incidence, contamination and the management at field level, after harvest, consumers level and exporting stage (2) creating awareness about the international standards for exporting commodities and the possible economic consequences if failed to meet such standards (3) facilitating further research on the fate of mycotoxins in food chain (4) blending of traditional and modern scientific knowledge to develop effective methods of management of the mycotoxin problem.

References

Acharya, K.T. 1986. Oil cakes in India. II. Trade and utilization. *J. of the Oil Technol. Assoc. of India*. **18**:96–107.

Ahmad, S.K., and P.L. Singh. 1991. Mycofloral changes and aflatoxin contamination in stored chickpea seeds. *Food Additives and Contaminants*, **8**:723–730.

Anisa Athar, M., H.S. Prakash, H.S. Shetty. 1988. Mycoflora of Indian spices with special reference to aflatoxin producing isolates of *Aspergillus flavus*. *Ind. J. of Microbiol.*, **28**:125–127.

Anjana Singh. 1983. Mycotoxin contamination in dry fruits and spices. *Mycotoxins in Food and Feed*, edited by K.S. Bilgrami, Allied Press, Bhagalpur, India, pp. 55–68.

Ansari A.A., and A.K. Shrivastava. 1990. Natural occurrence of *Alternaria* mycotoxins in sorghum and ragi from North Bihar, India. *Food Addit. Contam.*, **7**:815–820.

Augustine, B., M. Parvathi and Indira Kalyanasundaram. 1984. Potential mycotoxins in wheat and wheat products. *J. of Food Sci and Technol.* India, **21**:312–316.

Balachandran, C. and K.R. Parthasarathy. 1996. Occurrence of cyclopiazonic acid in feeds and feedstuffs in Tamilnadu, India. *Mycopathologia*, **133**:159–162.

Balachandran, C. and R. Ramakrishnan. 1987. Influence of dietary aflatoxin on growth rate and hematology in broiler chickens. *Indian J. Poultry Sci.* **22**:105–108.

Basappa S.C., Jayaraman, V. Sreenivasamurthy and H.A.B. Parpia. 1967. Effect of B-group vitamins and ethyl alcohol on aflatoxin production by *Aspergillus oryzae* L. *Indian J. Exp. Biol.* **5**:262–263.

Basappa, S.C. and T. Shantha. 1996. Methods of detoxification of aflatoxins in foods and feeds-a critical appraisal. *J. of Food Sci. and Technol.*, India, **33**:95–107.

Begum, F., and S. Samajpati. 2000, Mycotoxins production on rice, pulses and oilseeds. *Naturwissenschaften*, **87**:275–277.

Bhandari, P.C. and B. Bhandari. 1980. Aflatoxin and Indian childhood cirrhosis. *Indian Pediatr.*, **17**:593–596.

Bhat, R.V., P.H. Shetty, P. Amrith Rao and V. Sudershan. 1997a. A food borne disease outbreak due to the consumption of moldy sorghum and maize containing fumonisin mycotoxins. J. *Toxicol. Clinicol. Toxicol.*, **35**:249–255.

Bhat R.V., S. Vasanthi, B.S. Rao, R.N. Rao, V.S. Rao, K.V. Nagaraja, R.G. Bai, C.A. Prasad, S. Vanchinathan, R. Rooy, S. Saha, A. Mukherjee, P.K. Ghosh, S. Toteja and B.N. Saxena. 1996. Aflatoxin B_1 contamination in ground nut samples collected from different geographical regions of India: a multicentric study. *Food Addit. Contam.*, **13**:325–331.

Bhat, R.V., 1989b. Control of mycotoxin contamination in India *Lancet*, **2(8662)**:575–576.

Bhat, R.V., R.B. Sashidhar, Y. Ramakrishna and K.L. Munshi. 1989a. Outbreak of trichothecene mycotoxicosis associated with consumption of mould damaged wheat products in Kashmir Valley in India. *Appl. Environ. Microbiol.*, **55**:35–37.

Bhat, R.V., S.R. Beedu, Y. Ramakrishna and K.L. Munshi. 1989b. Outbreak of trichothecene mycotoxicosis associated with consumption of mould damaged wheat production in Kashmir Valley, India. *Lancet*, **1(8628)**:35–37.

Bhat, R.V. 1988. Mould deterioration of agricultural commodities during transit; problems faced by developing countries. *International J. of Food Microbiol.*, **7**:219–225.

Bhat, R.V. 1989a, Risk to human health associated with consumption of ground nuts contaminated with aflatoxins. In: *Proceedings of International workshop on Aflatoxin contamination of groundnuts*, 6–9th Oct., 1987, ICRISAT, India, pp. 19–29.

Bhat, R.V., V. Nagarajan and P.G. Tulpule. 1978. *Health hazards of mycotoxins in India*, Indian Council of Medical Research, New Delhi, India, p. 58.

Bhat, R.V., S. Vasanthi, B. Sashidhar Rao, R. Nageswara Rao, R.V. Sudarshan, K.V. Nagaraja, R. Girija bai, C.A. Krishna Prasad, S. Vanchinathan, R. Roy, S. Saha, A. Mukherjee, P.K. Ghosh, G.S. Toteja, B.N. Saxena. 1997. Aflatoxin B_1 contamination in regions of India — a multicentric study. *Food Addit. and Contam.*, **14**:151–156.

Bhavanishankar, T.N. and T. Shantha. 1987. Natural occurrence of *Fusarium* toxins in peanut, sorghum and maize from Mysore, *J. of Sci. and Food and Agric.*, **40**:327–332.

Bilgrami, K.S. 1984. Mycotoxins in Food. *J. Ind. Bot. Soc.*, **63**:109–120.

Bilgrami, K.S. 1985. Mycotosins in dry fruits and spices, In: *Mycotoxins in Food and Feed.* edited by K.S. Bilgrami, Allied Press, Bhagalpur, India, p. 29.

Bilgrami, K.S., Choudhary and A. Masood. 1991. Aflatoxin contamination in mustard *Brassica juncea* in relation to agronomic practices. *J. Sci. Food Agric.*, **54**:221–228.

Bilgrami, K.S. and R.S., Misra. 1982. Aflatoxin production by *Aspergillus flavus* in storage and standing maize crops. *Advancing Frontiers of Mycology and Plant Pathology*, Edited by K.S. Bilgrami, R.S. Misra and P.C. Misra, Today and Tomorrow, New Delhi, pp. 67–68.

Bilgrami, K.S., R.S. Misra and K.K. Sinha. 1979. Aflatoxin production and loss in calorific value of maize due to *Aspergillus parasiticus. Current Sci.*, **48**:642–643.

Bilgrami, K.S., R.S. Misra, K.K. Sinha and P.L. Singh. 1979. Prevention of aflatoxin production in liquid and solid substrates through some plant extracts. *Indian J. Exptl. Biol.*, **17**:1151.

Bilgrami, K.S., R.S. Misra, K.K. Sinha and P.L. Singh. 1980. Effect of some wild and medicinal plant extracts on aflatoxin production and growth of *Aspergillus flavus* in liquid cultures. *J. Ind. Bot. Soc.*, **59**:123–126.

Bilgrami, K.S., R.S. Misra, S.K. Sinha and P.L. Singh. 1981. Inhibition of aflatoxin production by ferulic acid on some cereals and oil seeds. *Current Sci.*, **50**:997–998.

Bilgrami, K.S., T. Prasad, R.S. Misra and K.K. Sinha. 1980a. Survey and study of mycotoxin producing fungi associated with grains in the standing maize crop. *ICAR Final report*, Bhagalpur University, Bhagalpur, India, p. 98.

Bilgrami, K.S., T. Prasad, R.S. Misra and K.K. Sinha, 1981a. Aflatoxin contamination in maize under field conditions. *Indian Phytopath.*, **34**:67–68.

Bilgrami, K.S., T. Prasad, R.S., Misra, K.K. Sinha. 1981b. Aflatoxin contamination of maize under field conditions. *Ind. Phytopath.*, **34**:67–68.

Bilgrami, K.S., T. Prasad, R.S. Misra, K.K. Sinha. 1981c. Changes in protein and amino acid contents of maize seeds infested with aflatoxin producing strain of *Aspergillus parasiticus*. *Seed Res.*, **9**:52–56.

Bilgrami, K.S., K.S. Ranjan and A. Masood. 1992. Influence of the cropping pattern on aflatoxin contamination in pre-harvest kharif monsoon maize crop (*Zea mays*). *J. of Sci. of Food and Agric.*, **58**:101–106.

Bilgrami, K.S., K.S. Ranjan and S.P. Sinha. 1986. Aflatoxin induced histopathology of skin in guinea pigs. *Proc. Nat. Acad. Sci.* India, **52**:280–283.

Bilgrami, K.S., A. Singh, K.S. Ranjan. 1984a. Detoxification of aflatoxin in dry fruits and spices through light and heat treatment. *Nat. Acad. Sci. Letters*, **7**:273–274.

Bilgrami K.S., K.K. Sinha. 1983. Aflatoxin contamination in agricultural commodities with special reference to maize and groundnut. *Phytopathol.*, **73**:253–263.

Bilgrami, K.S. and K.K. Sinha. 1984a. Mycotoxins in cereals. *Rev. Trop. Pl. Path.*, **1**:335–374.

Bilgrami, K.S. and K.K. Sinha. 1984b. Mycotoxin contamination in food and its control. *Ind. Rev. Life Sci.*, **4**:19–36.

Bilgrami, K S., K.K. Sinha and Premalatha Singh. 1982. Prevention of aflatoxin production on cereals and oil seeds by O-vanillin, *Current Sci.*, **51**:138.

Chakrabarti, D.K. 1987. A review of deterioration of oil seeds by fungi with special reference of India. *International Biodeterioration*, **23**:137–157.

Chakrabarti, D.K. and S. Ghosal. 1986. Occurrence of free and conjugated 12, 13-epoxytrichothecenes and zeralenone in banana fruits infected with *Fusarium moniliforme*. *Applied and Environ. Microbiol.*, **51**:217–219.

Chary, M.P. and S.M. Reddy. 1987b. A survey of mycotoxin producing fungi associated with dehusked rice grains. *Ind. Bot. Rep.*, **6**:101–103.

Chary, M.P. and S.M. Reddy. 1987a. Mycotoxins contamination of dehusked rice in the flooded areas of Warangal. *Nat. Acad. Sci. Letters*, **10**:129–132.

Chary, M.P., S. Giridham and S.M. Reddy. 1986. Influence of different seed borne fungi of rice on citrinin production by *P. citrinum*. *J. food Sci. and Technol.*, **23**:160–162.

Chary, P.M., and S.M. Reddy. 1987. Mycotoxins contamination of dehusked rice in the flooded areas of Warangal, *National Acad. of Science Letters*, **10**:129–132.

Chatterjee, D. and S.K. Mukherjee. 1994. Contamintion of Indian maize with fumonisin B1 and its effects on chicken macrophage. *Letters in Applied Microbiology*, **18**:251–253.

Chaturvedi K. and S.N. Chaturvedi. 1992. Mycoflora associated with different types of arecanut and its mixtures. *J. Ind. Bot. Soc.*, **71**:209–212.

Chenulu, V.V. and Dharam Vir. 1979. Seed mycoflroa, its role in grain spoilage and production of mycotoxins. *Bulletin of Grain Technol.*, **17**:148–157.

Choudhary, A.K. 1992. Influence of microbial co-inhabitants on aflatoxin synthesis of *Aspergillus flavus* on maize kernels. *Letters in Applied Microbiol.*, **14**:143–147.

Choudhary, A.K. and K.K. Sinha. 1993. Competition between a toxigenic *Aspergillus flavus* strain and other fungi on stored maize kernels. *J. of Stored Products Res.*, **29**:75–80.

Chourasia, H.K. 1993. Growth, sclerotia and aflatoxin production by *A. parasiticus* infleuence of food preservatives. *Lett. in Appl. Microbiol.*, **17**:204–207.

Cotty, P.J., P. Bayman, D.S. Egel and K.S. Elias. 1994. Agriculture, Aflatoxins and *Aspergillus*, In: *The Genus Aspergillus*, edited by Keith A. Powell, Plenum Press, NY, USA. pp. 1–27.

Damodaran, C., C.S. Ramdoss and E.R.M. Shanmugasundaram. 1973. A rapid procedure for the isolation, identification and estimation of citrinin. *Anal. Biochem.*, **52**:482–488.

Dandavate, M.S. and P.A. Shinde. 1986. Mycotoxin producing potential of *Aspergillus* sp. from red pepper. *J. of Mah. Agric. University*, **11**:271–273.

Devaraj, R.H., K.R. Shanmugasundaram and E.B. Shanmugasundaram. 1983. Changes in carbohydrate metabolism during patulin toxicosis studied in chicks. *J. Ind. Bot. Soc.*, **64**:65–71.

Dharmsilal, K. and S.P. Sinha. 1994. Effect of retinal on ochratoxin produced genotoxicity in mice. *Food and Chemical Toxicol.*, **32**:471–475.

Dhavan, A.S. and M.R. Choudary. 1995. Incidence of aflatoxins in animal feedstuffs: a decade's scenario in India. *J. AOAC. Int.* **78**:693–698.

Dhuley, J.N. 1997. Effect of some Indian herbs on macrophage functions in ochratoxin A treated mice. *J. Ethnopharmacol.*, **58**:15–20.

Dwivedi, P.K., R.P.S. Tyagi and P.C. Bansode. 1990. Studies on fungal and mycotoxin contamination of stored millets Kodon (*Papsalum scrobiculatum*) and Kulki (*Panicum miliare*). *J . of Food Sci and Technol.* India, **27**:111–112.

Fasiha Rehana and S.C., Basappa. 1990. Detoxification of an aflatoxin B_1 in maize by different cooking method. *J. of Food Sci and Technol.* India, **27**:327.

Gajendra Prasad, Niranjan Kumar and K.K. Sinha. 1993. Effect of aflatoxin B_1 on seed germination, seedling growth and chlorophyll contents of maize (var. suwan composite) *J. Ind. Bot. Soc.*, **72**:305–306.

Gambhir, S.P. and D.N. Khairnar. 1986. Mycoflora of certain oil seeds from Nasik area. *Geobios New Reports* 5, 190–192.

Gangopadhyay, S. and N.K. Chakrabarti. 1981. Mycotoxins in stored rice. *Curr. Sci.*, **50**:272–275.

Gayatri Murjani. 2000. Chronic aflatoxicosis in fish and its relevance to human health, In: *Shaping the future*, Global dialogue Expo. 2002, Aug. 29th–31st, 2002 supported by Volkswagen foundation, Hannover, Germany.

Gelderblom, W.C.A., K. Jasiewicz., W.F.o. Marasa, P.G. Thiel, R.M. Horak, R. Vleggar, N.P.J. Kriek. 1988. Fumonisins-novel Mycotoxins with cancer-promoting activity produced by *Fusarium moniliforme. Appl. Environ. Microbiol.*, **54**:1806–1811.

Ghewande, M.P. and G. Nagaraj. 1987. Prevention of aflatoxin contamination through some commercial chemical products and plant extracts in ground nut. *Mycotoxins Res.* **3**:19–24.

Gopal T. 1985. Mycotoxins, their effects on poultry. *Poultry Adviser*, **18**:41–44.

Ghosal, S., D.K. Chakrabarti and K.C.B. Chaudhary. 1977. The occurrence of 12,13-epoxytrichothecenes in seeds of safflower infected with *Fusarium oxysporum* f. sp. carthmi. *Experientia.* **33**:574–575.

Ghosal, S., D.K. Chakrabarti, A.K. Srivasatava and R.S. Srivastava. 1982. Toxic 12,13-epoxytrichothecenes from anise fruits infected with *Trichothencium roseum*, *J. of Agric. Food Chem.* **30**:106–109.

Ghosh, S.K., M. Desai G.L. Pandya and K. Venkaiah. 1997. Airborne aflatoxin in the grain processing industries in India. *Ind. Hyg. Assoc. J.*, **58**:583–586.

Giridhar, P., G. Laxma Reddy, S.M. Reddy and M. Ather. 2000. Toxicological studies of cyclopiazonic acid on Broiler chicks. *Ind. J. of Poultry Sci.*, **35**:45–49.

Giridhar, P., G. Orlich, E. Komor and S.M. Reddy. 2001. Elicitation of phytoalexins accumulation and phloem characteristics by some fungi and mycotoxins in *Ricinus communis. Advances in Plant Sciences*, **14**:211–224

Giridhar, P., M.V., Rajam and S.M. Reddy. 1999. Effect of Inhibitors of polyamine biosynthesis inhibitors on Aflatoxin B_1 and Ochratoxin A production. *Advances in Plant Sciences*, **12**:41–44.

Giridhar, P. and S.M. Reddy and M.V. Rajam. 1997. Contol of *Penicillium citrinum* growth and citrinin production by polyamine biosynthesis inhibitors.

Giridhar, P. and S.M. Reddy. 1999a. Incidence of mycotoxigenic fungi on cashewnut. *Journal of Mycology and Plant Pathology*, **28**:335–337.

Giridhar, P. and S.M. Reddy. 1996c. Inhibitory effect of black pepper on growth and toxin production of toxigenic fungi. *J. of Ind. Bot. Soc.*, **76**:43–45.

Giridhar, P. and S.M. Reddy. 1996. Effect of Some plant extracts on citrinin production by *Penicillium citrinum in vitro. Journal of Ind. Bot. Soc.*, **75**:153–154.

Giridhar, P. and S.M. Reddy. 1996a. Seed mycoflora in relation to mycotoxins of mustard and arecanut from A.P. *Geobios New Reports*, **16**:19–24.

Giridhar, P. and S.M. Reddy. 1996b. Mycoflora and mycotoxins of coconut copra collected from A.P. *J. of Ind. Bot. Soc.*, **75**:147–148.

Giridhar, P. and S.M. Reddy. 1997a. Incidence of Mycotoxin producers on spices from Andhra Pradesh. *J of Ind. Bot. Soc.*, **76**:161–164.

Giridhar, P. and S.M. Reddy. 1997b. Combined activity of temperature and water activity on biodeterioration and mycotoxin production by two storage moulds. *Advances in Plant sciences*, **10**:125–128.

Giridhar, P. and S.M. Reddy. 1997c. Effect of growth regulators on growth, citrinin production by *Penicillium citrinum. Nat. Acad. of Sci. Letters.* **20**:62–63.

Giridhar, P. and S.M. Reddy, 1997d. Effect of some food preservatives on seed mycoflora of fenugreek. *Seed Res.*, **25**:92–93.

Giridhar, P., and S.M. Reddy. 1997e. Influence of relative humidity on seed mycoflora in relation to mycotoxin production and seed health of black pepper (*Piper nigrum* L.) *Advances in Plant Sciences*, **10**:115–121.

Giridhar, P., and S.M. Reddy. 1997f. Influence of vitamins and chemical mutagens on citrinin production by *P. citrinum. Nat. Acad. of Sci. letters*, **20**:59–61.

Giridhar, P. and S.M. Reddy. 1997g. Efficacy of some Bacteria in the control of growth and citrinin production by *Penicillium citrinum. Ind. Phytopathol. Soc.*, **50**:1–5.

Giridhar, P. and S.M. Reddy. 1998. The effect of some polyamine biosynthesis inhibitors on mycotoxin production by *Aspergillus terreus. Nat. Acd. of Sci. Letters*, **21**:5–8.

Giridhar, P. and S.M. Reddy. 1999b. Mycoflora in relation to mycotoxins incidence on red pepper. *Advances in Plant Sciences*, **12**:85–88.

Giridhar, P. and S.M. Reddy. 1999c. Studies on mycotoxigenic fungi associated with fenugreek. *Advances in Plant Sciences*, **12**:249–254.

Giridhar, P. and S.M. Reddy. 1999d. Incidence of fungi in relation to mycotoxins on Garlic and Onion bulbs in Andhra Pradesh. *J. of Mycol. and Plant Pathol.*, **28**:347–349.

Giridhar, P. and S.M. Reddy. 2001a. Incidence of mycotoxigenic fungi on Raisins. *Advances in Plant Sciences*, **14**:291–294.

Giridhar, P. and S.M. Reddy. 2001b. Mycobiota and potential mycotoxins of Date fruit. *J. of Food Sci & Technol, India*, **38**:419–421.

Giridhar, P. and S.M. Reddy. 2001c. Phenolic antioxidants in the control of some mycotoxigenic fungi. *J. of Food Sci & Technol. India*, **38**:397–399.

Giridhar, P. and S.M. Reddy. 2002. Effect of storage structures on seed mycoflora and seed deterioration in black pepper. *Advances in Plant Sciences*, **15**:331–334.

Girish, G.K. and R.K. Goyal. 1986. Presence of mycoflora and mycotoxins in different food grains with particular reference to paddy/rice., *Bull. Grain Tech.* **24,** 157–177.

Girisham, S. and S.M. Reddy. 1989a. Efficacy of volatile compounds and food preservatives in the control of terreic acid production by *Aspergillus terreus. J. Ind. Bot., Soc.*, **68**:52–43.

Girisham, S. and S.M. Reddy. 1989b. Influence of relative humidity on biodeterioaration of pearl millet by *Aspergilus flavus. Geobios*, **16**:235–236.

Girisham, S. and S.M. Reddy. 1986. Seed mycoflora of pearl millet (*Pennisetum typhoides*) in relation to moisture, *Geobios New Reports*, **5**:121–125.

Girisham, S. and S.M. Reddy. 1988. Influence of humidity on biodeterioiration and patulin and terreic acid production by *Aspergillus terreus* in pearl millet. *J. Ind. Bot. Soc.*, **67**:59–60.

Girisham, S. and S.M. Reddy. 1987. Screening of different varieties of pearl millet for patulin and terreic acid production by *Aspergillus parasiticus. Ind. J. Mycol. and Plant Pathol.*, **17**:204–205.

Hareesh Vardhan Rao, P., P. Giridhar and S.M. Reddy. 1996a. Seed mycoflora in relation to seed variety. *Ind. Bot. Rep.*, **14**:7–9.

Hareesh Vardhan Rao, P., P. Giridhar and S.M. Reddy. 1996b. Effect of Volatile compounds on seed mycoflora of sorghum. *Ind. Phytopathol. Soc.*, **49**:41–44.

Hareesh Vardhan Rao, P., P. Giridhar and S.M. Reddy. 1996c. Influence of storage structures on seed mycoflora of sorghum. *J. of Ind. Bot. Soc.*, **75**:313–314.

Hareesh Vardhan Rao, P., P. Giridhar and S.M. Reddy. 2001a. Effect of culture filtrates of some seed borne fungi on seed germination and seedling growth of sorghum. *Proc. National Acad. of Sciences*, India, **71**:10–12

Hareesh Vardhan Rao, P., P. Giridhar and S.M. Reddy. 2001b. Fungal succession on sorghum seed harvested at different stages of maturity. *Indian Phytopathol.* **53(1)**:102–104.

Hussein, H.S. and M. Brasel. 2001. Toxicity, metabolism and impact of mycotxins on humans and animals. *Toxicology*, **167(2)**:101–134.

Janardhana, G.R., K.A. Raveesha and H.S. Shetty, 1999. Mycotoxin contamination of maize grains in Karnataka (India) *Food Chem. Toxicol.*, **37**:863–868.

Jayaraman, P. and I. Kalyanasundaram. 1990. Natural occurrence of toxigenic fungi and mycotoxins in rice bran. *Mycopathologia*, **110**:81–85.

Jayashree T., J.P. Rao, C. Subramanyam. 1999. Regulation of aflatoxin production by Ca_2+/ calmodulin-dependent protein phosphorylation and dephosphorylation. *FEMS Microbiology letters*, **183**:215–219.

Jayashree, T., and C. Subramnyam. 1999. Anti aflatoxigenic activity of eugenol is due to inhibition of lipid peroxidation. *Letters in Applied Microbiol.*, **28**:179–183.

Joshi, B.C., S.K. Srivastava, P.K. Dwivedi and S.K. Prajapathi. 1987a. Microbial spoilage of flood affected stored wheat. *Bulletin of Grain Technol.*, **25**:85–87.

Joshi, B.G., S.K. Srivastava and S.K. Prajapati. 1987b. Studies on detection of mycoflora and mycotoxins in paddy harvested from rain affected crops in Haryana, *Bull. Grain Tech.*, **25(2)**:188–190.

Joshi, B.G., S.K. Srivastava and S.K. Prajapati. 1986b. Studies on detection of mycoflora and aflatoxin B_1 in stored soybean, *Bull. Grain Tech.*, **24**:206–210.

Jyothi Saxena, S.K., P.H. Neeta Pande and B.S. Mehrotra. 1988. Mycotoxin producing potentials of the fungi isolated from some common spices. *Proc. Nat. Acad. Sci. India*, **58**:427–430.

Kalra, D.S. and K.C. Bhatia. 1990. Degnala disease in buffaloes and cattle: epidemiological investigations. *J. Environ Pathol Toxicol. Oncol.*, **10**:132–135.

Kaur, H., S.K. Jand and D.V. Joshi. 1994. Comparison of different media for screening of toxigenic strains of moulds. *Ind. Veterinary Journal*, **71**:859–861.

Kavita-Waghray and M. Uma Reddy. 1996, Efficacy of a modified Buchner funnel in the decontamination of aflatoxin B_1, *Indian Food Industry*, **15**:37–39.

Krishna Kumari, C., Nusrath, M., and Reddy, B.N., 1984. Aflatoxin contamination of coconut products from Andhra Pradesh. *Indian phytopathol.* **37**, 284–287.

Krishna Reddy V., P. Giridhar, M. Surekha, S.M. Reddy and M.V. Rajam. 1997. Polyamine biosynthesis inhibitors in the control of mycotoxins. In: *Fungi in Biotechnology* (ed. Anil Prakash) CBS Publishers and Distributors, New Delhi. pp. 205–208.

Krishna Reddy, V. and S.M. Reddy. 1989. Influence of different seed borne fungi of maize (*Zea mays* L.) on the production of cyclopiazonic acid by *Penicillium griseofulvum*, *Indian J. MIcrobiol.*, **29**:57–60.

Krishna Reddy, V. and S.M. Reddy. 1990. Elaboration of mycotoxin by seed borne fungi of maize. *National Acad. Sci. Letters*, **13**:11–13.

Krishna Reddy, V. and S.M. Reddy. 1992. Cyclopiazonic acid production by *Penicillium griseofulvum* in relation to different cultivars of maize. *World J. of Microbiol. and Biotechnol.*, **8**:208–209.

Krishnamachari, K.A., R.V. Bhat, V. Nagarajan, T.B. Tilak and P.G. Tulpule. 1977. The problem of aflatoxic human disease in parts of India — epidemiological and ecological aspects. *Ann Nutr Aliment.*, **31**:991–996.

Krishnamachari, K.A.V.R., R.V. Bhat, V. Nagarajan and T.B.G. Tilak. 1975. Hepatitis due to aflatoxicosis. An outbreak in western India. *Lancet* **1**:1061–1063.

Kulkarni, J.R., D.N. Padule, P.A. Shinde and D.K. Salunkhe. 1986. Aflatoxins in copra. *J. of Maharashtra Agric. Nut.*, **11**:199–201.

Kumar, S. and G. Prasad. 1992. Effect of medicinal plant (*Andrographis paniculata*) extract on aflatoxin production and growth of *Aspergillus flavus*. *Letters in Applied Microbiol.*, **15**:131–132.

Kumar, S. and A.K. Roy. 1993. Occurrence of aflatoxin in some liver curative herbal medicines. *Lett. in Applied Microbiol.*, **17**:112–114.

Laxma Reddy, G. and S.M. Reddy. 1992. Chemical control of satratoxin production of by *Stachybotrys atra* through food preservatives. *National Acad. of Sci. Lett.*, **15**:175–177.

Laxma Reddy, G. and S.M. Reddy. 1992. Effect of water activity (a_w) on biodeterioraion and trichothencene production by some fungi in cattle feed. *National Acad. Sci. letters*, **15**:237–240.

Laxma Reddy, G. and S.M. Reddy. 1994. Incidence of trichothecene mycotoxin producing fungi in oil seed cakes. *Ind. J. of Anim. Nutri.*, **11**:101–106.

Laxma Reddy, G. and S.M. Reddy. 1992b, Chemical control of satratoxin production by *Stachybotrys atra* through food preservation, *Nat. Acad. Sci. Letters*, **15**:175–177.

Lewis, C.W., J.G. Anderson and J.E. Smith. 1994. Health related aspects of the Genus *Aspergillus*, In: *Biotechnology Hand Books. 7 Aspergillus*, edited by J.E. Smith, Plenum Press, NY, London, pp. 219–261.

Lillehoj, E.B. 1973. Feed sources and conditions conducive to production of aflatoxins, ochratoxins, fumonisin toxins and zearalenone. *J. Am. Vet. Med. Assoc.*, **163**:1281–1283.

Madhyastha, M. and R.V. Bhat. 1984. *Aspergillus parasiticus* and aflatoxin production on black and white pepper and the inhibitory action of their chemical constituent. *Appl. Environ. Microbiol.*, **48**:376–379.

Mall, G.P., H.M. Pateria and S.K. Chauhan. 1986. *Indian Phytopath.* **39**:409–413.

Mall, O.P., P. Mehta, S.C. Agarwal and V.C. Vora. 1979. A survey of fungal contaminants of field crops with a view to evaluate their ability to produce mycotoxins. I. Fungal contaminants of wheat and barley harvested during spring, 1975, *Ind. J. of Microbiol.*, **19**:118–122.

Manorama and R.S. Singh. 1995. Mycotoxins in milk and milk products — a review. *J. of Dairying Foods and Home Sci.*, **14**:101–107.

Marasas, W.F.O. 1996. Fumonisins: History, world wide occurrence and Impact. In Fumonisins in Food; L.S. Jackson, J.W. Devreies, L.B. Bullerman (eds.), Plenum Press: New York, pp. 1–17.

Misra, N. 1987. Mycotoxins in spices, Investigation on the natural occurrence of aflatoxins in *Coriandrum sativum* L. *J. of Food Sci and Technol.*, India, **24**:324–325.

Misra, N. and S. Batra. 1987. Mycotoxins in Fennel seed, *J. of Food Safety*, **8**:157–160.

Mishra, N.K. and S.K. Daradhiyar. 1991. Mold flora and aflatoxin contamination of stored and cooked samples of pearl millet in the Paharia tribal belt of Santhal Paragana Bihar, India. *Applied and Environmental Microbiol.*, **57**:1223–1226.

Mitra, J., E.R. Rati and R. Joseph. 2000. Biodegradation of aflatoxin B_1 in Idli, a traditional fermented food product in India. *Ind. J. of Microbiology*, **40(2)**:119–123.

Monteriro, P.V., K. Sudhindra Rao and V. Prakash. 1996. *In vitro* interaction of ground nut proteins with aflatoxin B_1. *J. of Food Sci. and Technol.*, India, **33**:27–31.

Mukherjee, K. and A.V. Lakshminarasimham. 1995. Aflatoxin contamination of sorghum seeds during storage under controlled conditions. *Zentralbl. Bakteriol.*, **282**:237–243.

Mukit, A. and M.S. Kwatra. 1989. Studies on prevalence of pathology of aflatoxicosis in Khaki Campbell ducks. *Ind. J. Poult. Sci.*, **24**:292–294.

Mukit, A., S. Goswami and M. Bhattacharya. 1994. Histopathological and histoenzyamtic studies on spontaneous chronic aflatoxicosis in ducks. *Ind. J. of Poultry Sci.*, **29**:72–74.

Nagaraj, G. and K. Kumar. 1986. Location variations in the aflatoxins content of some Virginia ground nut varieties. *J. of the Oil Technologists Assoc.*, India, **18**:89–91.

Narasimham, M.J. 1968. On the presence of aflatoxins in copra. *Hindusthan Antibiot. Bull.*, **11**:104–105.

Neelakanthan, S., G. Indira Jasmine, R. Balasaraswathi and T. Balasubramanian. 1984, Incidence of Penicillic acid in apples, *Madras Agri. Journal*, **71**:43–45.

Neelakanthan, S., T. Balasubramanian, R. Balasaraswathi, Indira G. Jasmine and R. Swaminathan. 1978. Detection of penicillic acid in foods. *J. of Food Sci and Technol.* India, pp. 15, 125–126.

Neelakanthan, S.R., R. Balasaraswathi, T. Balasubramaniyan and Indira Jasmin. 1982. Incidence of patulin in scented supari. *Madras Agric. J.*, **69**:326–328.

Neelakanthan, S.R., Swaminathan, R., Balasubramaniyan, T., and Indira Jasmin, 1978, Incidence of toxigenic fungi associated with poultry feed, *Ind. Poultry Gaz.*, **62**:40.

NIN. 1992. National Institute of Nutrition, Hyderabad, Annual Report 1991–92, p. 64.

NIN. 1998. National Institute of Nutrition, Hyderabad, Annual Report 1996–98, pp. 35–36.

Nusrath, M. and V. Ravi. 1984. Mycoflora, mycotoxins and toxigenic fungi from paddy. *Indian J Bot.* **7**, 94–97.

Pallavi, R.M.V., D. Ramana and R.B. Sashidhar. 1997. A simple test tube screening method for identification of aflatoxin producing strains of *Aspergillus* sp. using coconut milk agar medium. *World J. of Microbiol. and Biotechnol.* 13, 713–714.

Patel, U.D., P. Govindarajan and P.J. Dave. 1989. Inactivation of aflatoxin B_1 by using the synergistic effect of hydrogen peroxide and gamma radiation. *Applied and Environ. Microbiology*, **55**:465–467.

Patkar, K.L., C.M. Usha, H.S. Shetty, N. Paster and J. Lacey. 1993. Effect of spice essential oils on growth and aflatoxin B_1 production by *Aspergillus flavus. Letters in Applied Micriobiol.*, **17**:49–51.

Patkar, K.l., C.M. Usha, H.S. Shetty, N. Paster and J. Lacey. 1994, Effects of spice oil treatment of rice on moulding and mycotoxins contamination. *Crop Protection*, **13**:519–524.

Prakash, K., B. Umakantha, G. Devegowda, C.V. Gowda and G. Krishnappa. 2000. Effect of individual and combined supplementation of Vitamin A and selenium on the performance of broiler chicks feed ochratoxin A. *Ind. J. Animal Sci.*, **70**:197–199.

Prasad, B.N., B.K. Sinha, A.K. Sinha. 1987. Aflatoxigenic fungi isolated from fish and its public health importance. *Ind. J. of Comp. Microbiol. Immunol. and Inf. Dis.*, **8**:135–136.

Prasad, G., S.S. Sahay and A. Masood. 1994. Inhibition of aflatoxin biosynthesis by the extract of *Amorphophallus campanulatus* (OL) and calcium oxalate. *Letters in Applied Microbiol.* **18**:203– 205.

Pratapkumar, S.H., V.S. Rao, R.J. Paramkishan and R.V. Bhat. 1997. Disease outbreak in laying hens arising from the consumption of fumonisin contaminated food. *Br. Poult. Sci.* **38**:475–479.

Premlatha Singh. 1988. Aflatoxin production and changes in protein and caloric value of some cereals infected by *Aspergillus parasiticus. J. Ind. Bot. Soc.*, **67**:1–2.

Premalata Singh and S.K. Ahmed. 1989. Pre-storage aflatoxin contamination and *Aspergillus flavus* association with various growth stages of chickpea (*Cicer arietinum). Annals of Agriculutral Science* **34**:805–812.

Premlatha Singh, Sita Bhagat and S.K. Ahmad. 1990. Aflatoxin elaboration and nutritional deterioration in some pulse cultivars during infestation with *Aspergillus flavus. J . of Food Sci and Technol.* India, **27**:60–62.

Pruthi, J.S. 1978. Mycotoxins in foods and feeds-their detection, estimation, preventive and curative measures. *Bulletin of Grain Technol.*, **16**:51–68.

Raisuddin, S. and J.K. Misra. 1991. Aflatoxin in betel nut and its control by use of food preservatives. *Food Additives and Contaminants*, **8**:707–712.

Rajakumari Kausal, Ashwani Kumar and Surendra Singh. 1994. Natural contamination of toxigenic *Fusarium* species and zearalenone in some oil cakes. *J. Ind. Bot. Soc.*, **73**:131–134.

Rajendra Singh, D.S.R., B. Bhadraiah and P. Rama Rao. 1991. Mycoflora and afaltoxin contamination in stored seeds of bitter gourd and Foeniculum. *Ind.J.of Microbial. Ecol.*, **2**:133–135.

Rajendran, G., Anthoni Raj and N.G.C. Lenger. 1987. Effect of pressure parboiling on mycotoxins and aflatoxins. *Bulletin of Grain Technol.*, **25**:90–91.

Rajesh Chahota, R.C. Katoch, S.P. Singh, Subhash Verma, Arvind Mahajan and K.B. Nagpal. 2000. Investigation of a severe aflatoxicosis out break among chicken in Himachal Pradesh. *Ind. J. Animal Sci.* **70**:22–24.

Ramakrishna, Y. and R.V. Bhat. 1990. Minicolumn chromatography for the detection of sterigmatocystin in agricultural commodities. *Mycopathologia* **110**:153–155.

Ramakrishna, Y., R.V. Bhat and M. Ravindranath. 1989. Production of deoxynivalenol by *Fusarium* isolates from samples of wheat associated with a human mycotoxicosis outbreak and from sorghum cultivars. *Appl. Environ. Microbiol.*, **55**:2169–2620.

Ramakrishna, Y., R.V. Bhat and S. Vasanthi, 1990. Natural occurrence of mycotoxins in stable food in India. *J. Agric. Food Chem.*, **38**:1857–1859.

Ranjan, K.S., S.S. Sahay and A.K. Sinha. 1992, The influence of storage structures on aflatoxin contamination in wheat and mustard. *J. of Food Sci and Technol.*, India, **28**:221–224.

Rao, B.L. and A. Hussain, 1985. Presence of cyclopiazonic acid in kodo millet (*Papsalum scrobiculatum)* causing "Kodua Poisoning" in man and its production by associated fungi . *Mycopathologia*, **89**:177–180.

Rao, G.V., P.S. Rąo, S. Girisham and S.M. Reddy. 1985. A novel spray reagent for chromatographic detection of trichothecene toxins. *Curr. Sci.*, **54**:507–509.

Rathinavelu, A. and E.R.B. Shanmugasundaram. 1984. Simple colorimetric estimation of cyclopiazonic acid in contaminated food and feeds. *J. of the A.O.A.C.* **67**:38–40.

Rati, E.R., T. Shantha and H.P. Ramesh. 1991. Effect of long term feeding and withdrawl of aflatoxin B_1 and ochratoxin A on kidney cell transformation in albino rats. *Indian J. of Exp. Biol.*, **29**:813.

Rati, E.R. and T. Shantha. 1994. Incidence of aflatoxin in groundnut based snack products. *J. of Food Science and Technol.* India, **31**:327–329.

Rati, E.R., S.C. Basappa and V.S. Murthy. 1979. Studies on the occurrence of ochratoxins in food grains. *J. of Food Sci and Technol.*, India, **16**:113–114.

Rati, E.R., V. Prema and T. Shantha. 1987. Modification of Pon's method of estimating aflatoxin B_1 in corn, groundnut and ground nut cake. *J . of Food Sci. and Technol.* India, **24**:90–91.

Reddy, A.S. and S.M. Reddy. 1983. Elaboration of mycotoxins by fungi associated with til (*Seasamum indicum* L.) *Curr. Sci.*, pp. 613–614.

Reddy, B.N. and M. Nusrath. 1985. Mycotoxin contamination in stored sorghum *Curr. Sci.*, pp. 31–32.

Reddy, B.N. and M. Nusrath. 1986. Mycotoxin contamination in stored sorghum, *Science and Culture*, **52**:31–32.

Reddy, S.M., D. Rajakumari, and V. Krishna Reddy. 1997. Toxicological studies of gliotoxin production by *Trichoderma viridae. Ind. J. Poult. Sci.*, **32**:169–172.

Reddy, S.V., D.K. Mayi, M.U. Reddy, K. Thirumal Devi and D.V. Reddy. 2001. Aflatoxins B_1 in different grades of chillies (*Capsicum annuum* L.) in India as determined by indirect competitive–ELISA. *Food Addit Contam.*, **18**:553–558.

Reddy, V.K. and S.M. Reddy. 1990. Efficacy of certain food preservatives in the control of cyclopiazonic acid production by *Penicillium griseofulvum, J. of Food Sci and Technol.*, India, **27**:180–181.

Roy, A.K. and H.K. Chourasia. 1989. Effect of temperature on aflatoxin production in *Mucuna pruriens* seeds. *Applied and Environmental Microbiol.*, **55**:531–532.

Roy, A.K. and S. Kumar. 1993. Aflatoxin problem in medicinal seeds under storage. *Vistas in Seed Biology.* **I**:381–393.

Sahay, S.S. 1988. Occurrence of aflatoxin in mustard seeds, cakes and oil. *J. Ind. Bot. Soc.*, pp. 188–190.

Sahay, S.S. and T. Prasad. 1990. The occurrence of aflatoxins in mustard and mustard products. *Food Additiv Contam.*, **7**:509–513.

Sahay, S.S. and T. Prasad. 1990. The occurrence of aflatoxins in mustard and mustard products. *Food Additives and Contaminants*, **7**:509–513.

Sahay, S.S., T. Prasad and K.K. Sinha. 1990. Post-harvest incidence of aflatoxins in *Linum usitatissimum* seeds. *J. of Sci. of Food and Agric.*, **53**:169–174.

Sargeant, K., A. Sheridan, J. O'keliey and R.B.A. Carnaghan. 1961. Toxicity associated with samples of groundnut. *Nature*, **192**:1096–97.

Sashidhar, R.B. 1993. Fate of aflatoxin B_1 during the industrial production of edible defatted peanut protein flour form raw peanuts. *Food Chem.* **48**:349–352.

Sashidhar, R.B., M.V.V. Murthy and R.V. Bhat. 1992b. Analytical quality assurance for analysis in agricultural export produce. *Food Chem.* **43**:403–405.

Sashidhar, R.B., R.V. Bhat and S. Vasanthi. 1989. Rapid detection of aflatoxin by pressure mini-column technique: a collaborative trial. *Curr. Science*, **58**:882–885.

Sashidhar R.B., V. Sudershan Rao, Y. Ramakrishna and R.V. Bhat. 1989. Rapid and specific method for screening ergosterol as an index of fungal contamination in cereal grains. *Food Chemistry*, **31**:51–56.

Sashidhar, R.B., Y. Ramakrishna and R.V. Bhat. 1988. Rapid detection of aflatoxins by pressure mini-column technique. *Analytical letters*, **21**:507–518.

Sashidhar, R.B., Y. Ramakrishna and R.V. Bhat. 1992a. Moulds and mycotoxins in sorghum stored in traditional containers in India. *J. of Stored products Res.*, **28**:257–260.

Saxena, J. and B.S. Mehrotra. 1990. The occurrence of mycotoxins in some dry fruits retail marketed in Nainital District of India. *Acta Alimentaria*, **19**:221–224.

Shank, R.C., G.N. Wogan, J.B. Gibson and A. Nondasuta. 1972. Dietary aflatoxins and human liver cancer II: Aflatoxins in market foods and food stuffs of Thailand and Hong Kong . *Food Cosmet Toxicol.*, **10**:61.

Shantha T. 1999. Fungal degradation of Aflatoxin B_1. *Natural Toxins*, **7**:175–178.

Shantha, T. and E.R. Rati. 1990. Isolation and characterization of aflatoxin inhibitory metabolite from *A. niger*, *Current Sci.*, **59**:326.

Shantha, T., E.R. Rati and R. Joseph. 1996. Reversal of growth inhibition of *Bacillus megaterium* due to aflatoxin by Coffee and tea extracts. *Letters in Applied Microbiology* **23**:437–438.

Shantha, T. and V. Sreenivasa Murthy. 1981. Sunlight to partially detoxify groundnut (peanut) cake flour contaminated with aflatoxin B_1. *J. Assoc. Off. Anal. Chem.*, **164**:291–293.

Shantha, T., V. Sreenivasa Murthy, E.R. Rati and V. Prema. 1986. Detoxification of groundnut seeds by urea and sunlight. *J . of Food Safety*, **71**:225–231.

Sharma, Y.P. and G. Sumbali. 1999. Incidence of aflatoxin producing strains and aflatoxin contamination in dry fruit slices of quinces (*Cydonia oblonga* Mill.) from the Indian state of Jammu and Kashmir. *Mycopathologia*, **148**:103–107.

Sharma, A. and S.R. Padwal Desai. 1989. A note on aflatoxin biogenesis in the presence of benomyl and carbendazim. *J . of Food Sci and Technol.* India, **26**:366–367.

Sharma, A., S.R. Padwal Desai and G.B. Nadkarni. 1988. Effect of aldehyde trapping agents on ethylene and aflatoxin biogenesis in *Aspergillus parasiticus*. *J. of Agri. and Food Chem.* **36**:546–548.

Shetty, P.K. and R.V. Bhat 1997a. Natural occurrence of fumonisins B_1 and its co-occurrence with aflatoxin B_1 in Indian sorghum, maize and poultry feeds.

Shetty, P.K. and R.V. Bhat. 1997b. Differential production of fumoninsin B1 in maize inbreds and hybrids in laboratory. *Cereal Res. Communications*, **25**:1011–1015.

Shetty, P.K., V.S. Rao, R.J. Paramkishan, R.V. Bhat. 1997. Disease outbreak in laying hens arising from the consumption of fumonisins contaminated food. *Br. Poultry Sci.*, **38**:475–479.

Shrivastava, A.K., K.S. Ranjan and A.A. Ansari. 1991. Inhibition of aflatoxin biosynthesis by light in liquid culture and food grains. *J. of Food Sci. and Technol.* India, **28**:189–190.

Singh, A., M.S. Oberoi, S.K. and B. Singh. 1994. Epidemology of inclusion body hepatitis in poultry in northern India from 1990 to 1994. *Rev. Sci. Tech.* **15**:1053–1060.

Singh, S.N. and L. Chand. 1993. Inhibition of aflatoxin production by garlic extract and sodium bicarbonate. *Crop Research*, **6**:149–154.

Singh, T., R.P.S. Tyagi and B.K. Varma. 1982. Study of the occurrence of aflatoxin B_1 in food grains. *J. Food. Sci. & Tech.*, **19**:35–37.

Singh, T., R.P.S. Tyagi, S.K. Srivastava, and G.K. Girish. 1983. Microorganisms in stored food grains and their activities. *Bull. Grain Tech.*, **21**:63–73.

Singh, V., V. Singh. 1998. Enzymatic degradation of aflatoxins in shake culures of *Aspergillus flavus*. CMI 102566. A possible role of peroxide in biological control mechanism. *J. of Phytopathol Res.*, **11**:7–10.

Sinha, A.K., K.S. Bilgrami, K.S. Ranjan and T. Prasad. 1988. Incidence of aflatoxins in mustard crop in Bihar, *Ind. Phytopathol.* **41**:434–437.

Sinha, A.K. and K.S. Ranjan. 1989. BGY-fluorescence and aflatoxin contamination in insect damaged maize crop. *Ind. Phytopathol.*, **42**:514–518.

Sinha, A.K. and K.S. Ranjan. 1991. A report of mycotoxin contamination in Bhutanese cheese. *J. of Food Sci. and Technol.* India, **28**:398–399.

Sinha, B.K., K. Ranjan and J.N. Pandey. 1999. Aflatoxin contamination of animal fed in Bihar. *Ind. J. of Veterinary Research*, **8**:31–38.

Sinha, B.K., K.S. Ranjan, T.N. Pandey. 2002. Mycotoxins, mycotoxigenic fungi and the biochemical changes in makhana (*Euryale ferox* Salisb) puffs of North Bihar, *J. of Food Sci. and Technol.*, **39**:38–41.

Sinha, K.K. 1987. Aflatoxin contamination of maize in flooded areas of Bhagalpur. *Appl. Env. Microbiol.*, **53**:1391–1393.

Sinha, K.K. 1988. Mycotoxin problem in maize crop in North Bihar, India, *Food Lab. Newsletter*, **14**:25–26.

Sinha, K.K. 1990. Incidence of mycotoxins in maize grains in Bihar State, India. *Food Addit. Contam.*, **7**:55–61.

Sinha, K.K. and A. Singh. 1987. Chemical changes in apples due to aflatoxin producing aspergilli. *J . of Food Sci. and Technol.*, India, **24**:4–45.

Sinha, K.K. and A.K. Sinha. 1991. Monitoring and identification of aflatoxins in wheat, gram and maize flours in Bihar state (India), *Food Addit. Contam.*, **8**:453–457.

Sinha, K.K., A.K. Sinha. 1992. Impact of stored grain pests on seed deterioration and aflatoxin contamination in maize *Journal of Stored Products Research*, **28**:211–219.

Sinha, K.K., A.K. Sinha, Gajendra Prasad. 1993. The effects of clove and cinnamon oils on growth and aflatoxin production by *Aspergillus flavus*. *Letters in Applied Microbiol.*, **16**:114–117.

Sivaswamy, S.N., B. Balachandran and S. Balanehru. 1991. South Indian Foods: contaminants and their effects. *Bull Environ. Contam. Toxicol.*, **47**:251–260.

Sreenivasamurthy, N., H.A.B. Parpia, S. Srikanta and A. Shankar Murthy. 1967. Detoxification of aflatoxin in peanut meal by hydrogen peroxide. *J. Assoc. Off. Anal. Chem.*, **50**:350–354.

Srivastava, J.L.1987. Mycotoxin problems in food in India. Paper presented at the Joint FAO/WHOIUNDP Second International Conference on Mycotoxins held at Bangkok, Thailand from September 28–October 3, 1987.

Srivastava, P.K. 1986. Controlling mold damage in grains during storage. *Bull Grain Tech.*, **24**:64–69 .

Subramaniyan, P. and A.S. Rao. 1974. Occurrence of aflatoxins and citrinin in ground nut (*Arachis hypogaea* L.) at harvest in relation to pod condition and kernel moisture content. *Curr. Sci.*, **43**:707–710.

Subramaniyan, T. 1982. Colorimetric determination of patulin produced by *Penicillium patulum*. *J. Assoc. Off. Anal.Cehm.*, **65**:5–7.

Subramaniyan, T., M.N. Kuppuswamy and E.R.B. Shanmugasundarm. 1978. Colorimetric determination of terreic acid produced by *Aspergillus terreus* . J. *Asscoc. Off. Anal. Chem.*, **61**:581–583.

Sudershan, R.V., G.S. Prasad, T.P. Krishan and R.V. Bhat. 1992. Field level evaluation of aflatoxin detection kit. *J. of Food Protection*, **55**:392–394.

Sudhakar Rao G.V.S. and T.V. George. 2000. Production of sterigmatocystin on rice grains by *Aspergillus versicolor*. *Ind. J. Animal Sci.*, **70**:759–760.

Suman Mor and Kishan Singh. 2000. Incidence of toxigenic molds in dairy foods and animal feeds. *Ind. J. Animal Sci.*, **70**:766–768.

Suman Rai and S.N. Chaturvedi. 1993. Aflatoxin contamination in chewing tobacco (*Nicotiana tabaccum* L.). *J. Indian Bot. Soc.*, **72**:189–190.

Surekha, M. and S.M. Reddy. 1990. Succession of mycoflora in groundnut fodders, *Animal Feed Sci. and Technol.*, **31**:167–171.

Surekha, M. and S.M. Reddy. 1991. Mycotoxigenic moulds in sorghum fodder. *Ind. J. of Anim. Nutr.*, **8**:217–220.

Surekha M. and S.M. Reddy 1992. Efficacy of food preservatives in the control of pentriem B production by *Penicillium aurantiogriseum*. *J. Food Sci. Technol.* **6**:13.

Surekha, M. and S.M. Reddy. 1992. Efficacy of food preservatives in the control of Penitrem A production by *Penicillium aurantiogriseum*. *J. of Food Sci and Technol.* India, **29**:397–398.

Thirumala Devi, K., Mayo Ma, G. Reddy, K.E. Emmanuel, Y. Larondelle, D.V. Reddy. 2001. Occurrence of ochratoxin A in black pepper, coriander, ginger, and turmeric in India. *Food Addit Contam.*, **18**:830–835.

Tripathi, R.K. 1973. Aflatoxins in sorghum grain moulds infected with head moulds. *Ind. J. Expt. Biol.*, **11**:361–362.

Tyagi, R.P.S., T. Singh and S.K. Srivastava. 1986. Studies on detection of mycoflora and aflatoxin B1 in rice under prolonged storage, *Bull. Grain Tech.*, **24**:58–36.

Usha, C.M., K.L. Patkar, H.S. Shetty, R. Kennedy and J. Lacey. 1994. Mycoflora of developing sorghum grains with special reference to *Aspergillus flavus*. *Tropical Science.*, **34**:353–360.

Vasan, B.S. 1980. Handling of high moisture paddy wet season a practical approach to the existing problems. *Bull. Grain Tech.*, **18**:223–232.

Vasanthi, S. and R.V. Bhat. 1997. Aflatoxin intake from maize-based diets in a rural population in southern India. *J. of Sci. of Food and Agric.*, **73**:226–230.

Vasanthi, S., R.V. Bhat and G. Subbulakshmi. 1997. Aflatoxin intake from maize — based diets in a rural population in southern India. *J. Sci. Food Agric.*, **73**:226–230.

Verma, S.H. and S. Pandey. 1987. Effect of aflatoxin B_1 on the Haematopoictic tissue of air-breathing teleost *Cahanna punctatus*. *J. Zool. Scoc.*, India, **40**:79–87.

Verma, S.H., S. Pandey and R.A.B. Verma. 1988. Effect of aflatoxin B_1 on the liver of an air breathing teleost, *Channna punctatus*. *National Symposium on Environemtnal Importance on Animals and Aquaculture.* University of Kalyani, M.P., India.

Vijayalakshmi, M. and A.S. Rao. 1985. Toxigenic fungi associated with sunflower seeds during development. In: *Trichothecenes and other Mycotoxins* J. Lacey (ed.), John Wiley & Sons Ltd. Chap., **4**:33–45.

Vora, V.G. 1978. A survey of toxin producing fungi and mycotoxins associated with post harvest deterioration of field crops grown for human and animal consumption. *Final Report (PL 480 scheme)*, CDRI, Lucknow, India.

Waghray, S., C.S. Reddy and A.P.K. Reddy. 1988. Seed mycoflora and aflatoxin production in rice. *Indian Phytopathol.*, **41**:92–494.

Yadgiri, B. and P.G. Tulpule. 1974. Aflatoxin in Buffalo milk. *Ind. J. Dairy Sci.*, **27**:293–297.

Botanical Pesticides — A Friendly Natural Alternative

Vadde Ramakrishna and *Ch. Narayanaiah**

The indiscriminate use of chemical pesticides has given rise to many well-known and serious problems, including genetic resistance of pest species, toxic residues in stored products, increasing costs of application, hazards from handling etc. The problems caused by pesticides and their residues have increased the need for effective, biodegradable pesticides with greater selectivity. Agricultural entomologists, nematologists, and pathologists the world over are now actively engaged in research into the use of plants to fight agricultural pests and diseases, and to reduce the losses caused by them (WHO 1990; Ignacimuthu, 2004; Sharma and Sharma, 2004).

The plant world comprises a rich storehouse of biochemicals that could be tapped for use as pesticides. The toxic constituents present in the plant represent the secondary metabolites and have only an insignificant role in primary physiological processes in plants that synthesize them (Cooper and Johnson, 1984). Plants are the richest source of renewable bioactive organic chemicals. The total number of plant chemicals may exceed 4,00,000: of these 10,000 are secondary metabolites whose major role in the plants is reportedly defensive (Swain 1977). Numerous defensive chemicals belonging to various categories (terpenoids, alkaloids, glycosides, phenols, tannins, etc.), which cause behavioral and physiological effects on pests, have already been identified. These secondary compounds have usually been regarded as a part of the plants' defense against plant-feeding insects and other herbivores (Rosenthal and Janzen, 1979). Botanical insecticides tend to have broad-spectrum activity are safe and relatively specific in their mode of action and easy to process and use. Therefore,

* Department of Biotechnology, Sri Krishnadevaraya University, Anantapur-515 003. Andhra Pradesh, India.

they are considered relatively environmentally safe and less likely to kill natural enemies than insecticides with longer residual activity. Nonetheless, in addition to pest insects, some natural enemies may be killed by botanicals (Ignacimuthu, 2004). Farmers and small-scale industries can often easily produce botanical insecticides.

A few Commonly used Botanicals will be Briefly Described below

Annona

The most useful for making botanical insecticides are custard apple (*Annona muricata*), sweetsop (*Annona cherimola*) and soursop (*Annona muricata*). The botanical insecticide is contained in the unripe fruits, leaves, roots, and especially seeds. It works as a stomach poison, contact poison, repellant, and antifeedant. It is said to be effective against leafhoppers and caterpillars. To prepare, pound the seeds into powder and mix with soapy water. Strain the water before spraying (Parmer and Devkumar, 1993; Pawar and Singh, 1993).

Neem

Neem is an extract derived from the neem tree *(Azadiracta indica)* of semi-arid tropical regions, contains many active compounds that act as feeding deterrents and as insect growth regulators. First understand the ecosystem and how botanicals affect it! Among 231 compounds the main active ingredient is *azadiractin*, which is said to be effective on 200 types of insects, mites and nematodes. These include caterpillars, thrips and white-flies. It has low toxicity to mammals, including people. Both the juices from neem leaves, and the oil from the seeds, can be used to control pests. Neem interferes with the hormonal system in insects. It causes insects to moult improperly, thus killing them. Neem works best against chewing insects (Schmutterer and Ascher, 1987; Schmutterer *et al.*, 1990; National Research Council, 1992; Locke *et al.*, 1994; Quarles, 1994; Bhunde *et al.*, 1999; Stark, 2001; Koul *et al.*, 2003).

Neem Biopesticide (Emulsifiable Concentrate) is well suited for an "Integrated Pest Management" (IPM) Program because of the following salient features:

1. Neem Pesticide is a natural product, absolutely non-toxic, 100% biodegradable and environmentally friendly.
2. It is suited for mixing with other synthetic pesticides and in fact enhances their action.
3. None or lesser quantity of synthetic pesticides needs to be used, thereby reducing the environmental load.

4. Several synthetic pesticides being single chemical compounds cause easy development of resistant species of pests. Neem consists of several compounds hence development of resistance is impossible.
5. Neem does not destroy natural predators and parasites of pests thereby allowing these natural enemies to keep a check on the pest population

Pyrethrins

Pyrethrins are botanical insecticides made from the dried flower heads of Chrysanthemum flower, called Pyrethrum. Pyrethrins are nerve poisons that cause immediate paralysis to most insects. Low doses do not kill but have a "knock down" effect. Stronger doses kill. Toxicologists have intensively studied pyrethrins, therefore their Lethal Doses (LD_{50}s) for various target and non-target species are well known. Pyrethrins are effective as broad-spectrum insecticides to control pests such as aphids, whiteflies, stinkbugs and mites. They breakdown so quickly after application, that they are widely considered safe for use on fruits ready for harvest (Wain, 1977).

Nicotine

Nicotine is a natural insecticide from *Nicotiana spp.* (tobacco) stems and leaves. It is extremely toxic and fast acting on most animals, including livestock such as cows and chicken. It is very toxic to insects as a contact insecticide. On per dose basis it is 30 times as toxic as rotenone. Useful for knocking down chafers and other leaf-eating insects in an emergency. An additional danger of using tobacco leaf extract is that this extract may contain a virus disease called Tobacco Mosaic Virus, or TMV. This virus disease affects a wide range of plants, mainly solanaceous crops like tomato and eggplant. When spraying tobacco extract, you probably are applying TMV as well! Nicotine kills insects by contact, and if inhaled or eaten. The most common use is to control soft-bodied insects such as aphids, mites, and caterpillars (Marino-Bettolo, 1977; Nakanishi, 1977).

Rotenone

Rotenone is extracted from the roots of several shrubs or vines that make bean-like seeds (legumes). Rotenone is found in tropical plants such as Jewel Vine (*Derris* spp.), Lacepod (*Lanchocarpus* spp.), and hoary pea (*Tephrosia* spp.). Rotenone is a contact and stomach poison. For tea, it can be used against caterpillars, aphids, and beetles. It is also toxic to fish.

Marigolds are often grown in gardens for their attractive flowers, and are even cultivated commercially for use as cut flowers. In addition, marigold can have a repellant effect on insects and nematodes. In Kenya, for example, dried marigold when incorporated into the nursery soil was found to improve overall seedling health. Other experiments showed that fresh marigold repels caterpillars on cabbage, but only for a few hours.

Chilli or chillipepper: the ripe fruits and seeds contain insecticidal compounds. Good results can be obtained in control of aphids and caterpillars. In experiments in Kenya on vegetables, highly concentrated chilli solution treatments produced the same number of marketable leaves.

Garlic has been long known for its insecticidal activity. It may also help control some plant diseases. Garlic contains garlic oil and allicine, which have insecticidal and bacterial effect. It can be extracted two ways: (1) Extract in water. For example, mash 100 garlic cloves in 0.5 l water and a little soap, and allow soaking for 24 hours. Filter before using. (2) Extract in oil (some say this gives a stronger extract). For example, grate 100 g of garlic and mix with 2 teaspoons of vegetable oil. Allow soaking for 24 hours. Then, dissolve 10 g of soap in 0.5 liters of water. Mix the soapy water, the oil, and the garlic. Stir well, and then filter through a fine cloth. Before using, mix with 10 liters of water. The price of garlic may make these recipes expensive.

Sweet basil, *O. basilicum* (syn. *americanum*) was reported by Irvine (1955) for its potential as a pesticide. Leaves and seeds are rich in essential oils, which are repellent, toxic or growth inhibitory to many insects (Grainge and Ahmed, 1988). In field and laboratory experiments treatments with seed extracts against bugs in turnip, *Bagrada cruciferarum,* the American bollworm, *Helicoverpa armigera* and aphids, *Aphis gossypii* increased mortality between 50–90% (Pandey and Verma 1982; Pandey *et al.*, 1983a; 1983b). There are other wild *Ocimum* species such as *O. suave* and *O. canum,* which have been used traditionally against pests. Branches of *O. suave* are often placed around windows and doors to keep the mosquitoes away. Chemical analyses of the essential oils from the plant revealed that the principal compound was eugenol, which was shown to have a strong repellent effect on mosquitoes (Chogo and Crank, 1981). Linalool is another terpenoid found in hoary basil, *O. canum*. This chemical compound is responsible for the toxic effect of dried leaves to the bruchid *Zabrotes subfasciatus* and other storage pests (Weaver *et al.*, 1991). Extracts from *O. suave* have shown promising results in bioassays on maize weevils, *Sitophilus zeamais,* carried out at ICIPE, Kenya. The plant was going to be in focus for future evaluation of botanical pesticides in stored grain at the village level (Hassanali and Lwande, 1989). Eugenol and other isolates of essential oils have also a strong inhibitory effect on the growth of many fungi (Garg and Siddiqui, 1992).

Sabadilla is an insecticide produced by grinding the seeds of the sabadilla plant, *Schoenocaulon officinale.* Sabadilla is both a contact and stomach poison and has shown greatest promise against several of the "true bugs," such as squash bug, chinch bug, harlequin bug and stink bugs. It has proven effectiveness against leaf-feeding caterpillars, Mexican bean beetles and thrips. The EPA permits use of sabadilla on certain vegetables, including squash, cucumbers, melons, beans, turnips, mustard, collards, cabbage, peanuts and potatoes. Sabadilla dusts can be highly irritating to the respiratory tract, often provoking a violent sneezing reaction if inhaled. Be sure to wear a dust mask when applying it and, as with all pesticides, follow precautions listed on the product labels. In addition, several of the alkaloids in sabadilla can cause rapid depression of blood pressure in mammals.

Nematicidal Property

Some plants like *Tagetes*, garlic, lemon grass and *Crotolaria* sps. have been widely used in inter cropping and have shown promising nematicidal property. When lemon grass was planted in suburbs in Rio de Janeiro, human reinfection by nematodes of the Ancyclostomadae in paddy fields occurred in only 12% of the population in one year and a half compared with 50% reinfection observed in similar period in an adjacent unplanted area. A serious study of the intercropping with the lemon grass, *Vetiver* or one of the mints in potato, tomato and other nematode susceptible cultivation would seem to be justified as an alternative to the use of synthetic nematicides. The intercrop plants would themselves be commercially exploitable (Gill and Lewis, 1971). Neem cake applied at the rate of 1225 kg/ha is reported to be effective against *Meloidogyne incognita* in tomato and egg plants (Gebreyesus and Chapya 1983). The treatment improved the yield without ill effect on quality. Besides cake, other parts of the neem tree have also been reported to affect nematodes. Incorporation of leaves of *Chrysanthemum* sps., *Melia azadarach, Tagetes patula, Datura fastuosa* and *Nerium indicum* reduced the incidence of *Meloidogyne* sps. after six week of decomposition (Gupta *et al.*, 1991)

Insect Growth Regulator

The insect growth regulators (IGRs) are generally selective in their toxicity, slow in their killing action and safe to both the environment and the natural enemies of insect pests. This is the outcome of biorational approaches to exploit physiological and biochemical processes unique to insects or arthopods with the aim of providing biochemical leads for the discovery of unique new chemical control agents. In agriculture and forestry, IGRs are generally compatible with the agroecosystem and possess great potential in the integrated pest management programme. The rational application of these materials may lead to new strategies for the prevention and delay of occurrence of insect resistance problem. Plants offer a rich source of novel compounds with IGR properties. The ample potential in using phytoecdysones as IGRs cannot be overlooked. It has been reported that there are more than 1,000 species of plants (belonging to more than 200 families, 800 genera) containing these bioactive substances. Isolation of Ajugarins from *Ajuga remota* is a good example (Foon, 1987). Research is now underway in China to formulate enriched products from plants with high IGR properties for practical application in the field. The chemistry and mechanism of toxic action of these plant products to insects are also being studied.

Juvenile Hormone

If insects undergoing metamorphosis are treated with juvenile hormone (JH), they moult into intermediates, which are half immature and half adult, which die without gaining reproductive competence.

JH and analogues are also present in some plants. One of the first observations was made by Carol Williams (1956) who noted that the water of Rio Negro, was rich in suspended organic materials from wood decay and free from mosquito larvae, as do other rivers of the Amazon basin. JH analogues have been found in various plants and it may interfere to some extent with the control of several species of insects.

There have been numerous reports on the occurrence of plant metabolites having JH activity, i.e. phytojuvenoids such as juvabione, farnesol, juvocimene I and II, sesamin and sesamolin, thujic acid, sterculic acid, tagetone, ostruthin, echinolone, baku-chiol, juvadecene. The first JH analogue was discovered by chance from paper and was termed as paper factor (third generation pesticides). The compounds isolated from paper made from Balsam fir (*Abies balsamea*) were identified as juvabione and dehydrojuvabione (Bower et al., 1966, 1976; Eisner, 1964; Stone, 1992).

Ryania

Ryania is an extract from the roots of *Ryania speciosa*, effective in controlling codling moth caterpillars, leaf eating beetles and thrips. It has relatively low toxicity to mammals. It also kills 'beneficial' insects as well as 'pests'. It breaks down fairly slowly. Therefore, it should not be used on fruits near harvest time.

Characteristics of Botanicals

Botanicals have been used for centuries and were widely used in the United States until the 1940s and 50s when synthetic pesticides were introduced. The synthetics quickly became popular because they did not break down as quickly. But insecticides that last longer can potentially leave residues in the soil and water supply, and on food. As with any insecticide, there are advantages and disadvantages associated with botanicals.

- Botanicals degrade rapidly from sunlight, air and proper moisture, which generally make them less toxic to the environment but may also require them to be applied more often, applied correctly, and with more precise timing.
- Botanicals act quickly to stop feeding of insect pests and often cause immediate paralysis or cessation of feeding, but they may not cause the insect's death for hours or days.
- Most botanicals have low to moderate toxicity to mammals, yet they are still poisons and pose a hazard to humans or to the environment.
- Most botanicals are not toxic to plants except insecticidal soaps and nicotine sulfate.
- Botanicals cost more than synthetic insecticides and may not be readily available.
- Potency of some botanicals may differ from one source or batch to the next.

Botanicals tend to be broad-spectrum insecticides, meaning they kill whatever they come in contact with—both "good" and "bad" insects.

Botanicals are processed into various forms: dusts and powders made from ground and dried plant parts; pure chemicals isolated from plants; and plant extracts or resins. These extracts and resins are formulated as liquid concentrates or impregnated onto dusts or wettable powders. Botanicals generally act in one of two ways: either as a contact poison when sprayed on the insect, or as a stomach poison when applied to the plant and eaten by the insect.

Acknowledgement

The authors are thankful to the Andhra Pradesh Netherlands Biotechnology Programme, IPE, Hyderabad, India for financial support.

Refecences

Bhonde S.B., S.G. Deshpande, R.N. Sharma. 1999. *In vitro* evaluation on inhibitory nature of some Neem formulations against plant pathogenic fungi. *Hindustan Antibiot Bull.*, **41(1–4)**:22–24.

Bowers, W.S., T. Ohta, J.S. Cleere and P.A. Marsella. 1976. *Plant Sci.*, **193**:542–547.

Bowers, W.S., M.J. Thompson and E.C. Uebel. 1966. *Science*, **5**:1020.

Chogo, J.B., and G. Crank. 1981. Chemical composition and biological activity of the Tanzanian plant *Ocimum suave*. *J. of Natural Products*, **44**:308–311.

Cooper, M.R. and A.W. Johnson. 1984. In *Reference Book 161*, Ministry of Agriculture, Fisheries and Food, London.

Cranshaw, W. 2004. Garden Insects of North America. Princeton University Press. Princeton.

Eisner, T. 1964. *Science*, **146**:1318.

Foon, C.S. 1987. In Proceedings of the 11th International Congress on Plant Protection, Manila, pp. 292–294.

Garg, S.C. and N. Siddiqui. 1992. Antifungal activity of some essential oils isolates. *Pharmazie*, **47**:467–468.

Gebreyesus, T. and A. Chapya. 1983. *In*: *Natural Products for Innovative Pest Management*. D.L. Whitehead and W.S. Bower (eds.), Pergamon Press, Oxford, pp. 237–241.

Gill, J.S. and C.T. Lewis. 1971. *Nature*, **232**:402.

Grainge, M. and S. Ahmed. 1988. Handbook of plants with pest-control properties. Resource systems institute, East-West center, Honolulu, Hawaii. John Wiley & Sons, New York.

Gupta, M., A. Bagchi and A.B. Ray. 1991. *J. Natural Product*, **54**:599–602.

Hassanali, A. and W. Lwande. 1989. In *Insecticides of Plant Origin* J.T. Arnason, B.J.R. Philogene and P. Morand (eds.), ACS Symp. Ser. 367, American Chemical Society, Washington DC, pp. 78–98.

Ignacimuthu, S. 2004. Green pesticides for insect management. *Current Science*, **86(8)**:1059–60.

Irvine, F.R. 1955. West African insecticides, *Colon. Pl. Anim. Prod.*, **5**:34–38.

Koul O., J.S. Multani, G. Singh, W.M. Daniewski, Berlozecki. 2003. 6β-hydroxygedunin from *Azadirachta indica*. Its potentiation effects with some non-azadirachtin limonoids in neem against lepidopteran larvae. *J. Agric. Food Chem.*, **751(10)**:2937–42.

Locke, J. 1994. Neem Oil Locks out Spores. *Agricultural Research*. June 1994.

Marini-Bettolo, G.B. (ed.) 1977. In *Natural Products and the Protection of Plants*, Elsevier Scientific Publishing Company, New York, pp. 5–16.

Nakanishi, K. 1977. *In*: *Natural Products and the Protection of Plants* G.B. Marini-Bettolo, (ed.), Elsevier Scientific Publishing Company, New York, pp. 185–210.

National Research Council. 1992. Neem — a tree for solving global problems. National Academy Press, Washington.

Quarles, W. 1994. Neem Tree Pesticides Protect Ornamental Plants. *The IPM Practitioner*, **16(10)**:1–13.

Pandey, U.K. and G.S. Verma. 1982. Effectiveness of some indigenous plant products against insect-pests of cruciferous vegetables, *Bagrada cruciferarum. Kirk. Z.ang. Ent.*, **69(2)**:129–132.

Pandey, U.K., A. Srivastava, C. Lekha and A. Singh. 1983a. Efficacy of certain plant extracts against brinjal aphid *Aphis gossypii* Glover. *Indian J. of Entomology*, **45**:313–314.

Pandey, U.K., A. Srivastava, A. Singh and M. Panday. 1983b. Evaluation of some plant origin insecticides against gram caterpillar, *Heliothis armigera* Hubn. *Indian J. Entomology*, **45**:211–212.

Parmar, B.S. and C. Devkumar (eds.), 1993. In *Botanical and Biopesticides*, West Vill Publishing House, New Delhi, pp. 197–199.

Pawar, A.D. and B. Singh. 1993. In *Botanical and Biopesticides* B.S. Parmar and C. Devkumar (eds.), West Vill Publishing House, New Delhi, pp. 188–196.

Rosenthal, G.A. and D.H. Janzen (eds). 1979. Herbivores: their interaction with secondary plant metabolites. Academic Press, New York.

Schmutterer, H. 1990. Properties and potential of natural pesticides from the neem tree, *Azadirachta indica*. *Annu. Rev. Entomol.*, **35**:271–97.

Schmutterer, H. and K.R.S. Ascher (eds.). 1987. Natural pesticides from the neem tree and other tropical plants. Proceedings from the third international neem conference, Nairobi, Kenya, 10–15 July 1986. GTZ, Eschborn, Germany.

Sharma, R.K. and S.D. Sharma. 2004. Green chemistry: A present day need for sustainable future. *Everyman's Science*, **34(2)**:93–99.

Stark J.D. 2001. Population-level effects of the neem insecticide, Neemix, on *Daphnia pulex*. *J. Environ. Sci. Health B.*, **36(4)**:457–65.

Stone, R. 1992. A Biopesticidal Tree Begins to Blossom. *Science*. February 28.

Swain, T. 1977. *Annu. Rev. Plant Physiol.*, **28**:479–501.

Wain, R.L. 1977. *In*: *Natural Products and the Protection of Plants*, Elsevier Scientific Publishing Company, New York, pp. 483–499.

Weaver, D.K., F.V. Dunkel, L. Ntezurubanza, L.L. Jackson and D.T. Stock. 1991. The efficacy of linalool, a major component of freshly milled *Ocimum canum* Sims (Lamiaceae), for protection against post harvest damage by certain stored product Coleoptera. *J. of Stored Products Research*, **27**:213–220.

WHO. 1990. Public health impact of pesticides used in agriculture. World Health Organization, Geneva, Switzerland.

Williams, C. 1956. *Nature*, **178**:212–213.

Applications of Biotechnology for Improvement of Vegetable Solanaceous Crops — A Review

Peddaboina Venkataiah, Thamidala Christopher, Mahesh Upender, Mamidala Praveen, Karampuri Subhash and *Nanna Rama Swamy**

ABSTRACT

Eggplant, pepper and tomato are major vegetable crops that have achieved tremendous popularity over the last century. They are grown in almost every country of the world. Development of protocols for *in vitro* selection can provide new advances for the production of agronomically useful cultivars. Attempts were also made to induce advetitiuos shoot regeneration, shoot elongation and plant regeneration using various concentrations and combinations of plant growth regulators and different explant types in all these crop plants. Micropropagation protocols have developed using meristematic explants. Somatic embryogenesis and plant regeneration methods have also been developed in these crop plants from diverse explants using different concetrations and combinations of plant growth regulators. Inview of the importance of somatic hybridization and androgenic hapoidy in vegetable breeding programme, the attempts have also been made to obtain plant regeneration from protoplast cultures and subsequent development of somatic hydrids, and anther and microspore cultures in these crop plants. The use of a combination of molecular and conventional breeding techniques could be the option for the development of cultivars resistant to biotic and abiotic stresses. This paper reviews the advances made in various aspects of tissue culture and genetic engineering approaches for the improvement of tomato, pepper and eggplant.

Key words: *Tomato, pepper, eggplant, regeneration, haploidy, somatic hybridization, transformation.*

* Plant Biotechnology Research Group, Department of Botany, Kakatiya University, Warangal-506 009, A.P., India. E-mail: swamynr_dr@yahoo.com

Introduction

Tomato (*Lycopersicon esculentum* Mill.), Eggplant (*Solanum melongena* L.) and Pepper (*Capsicum annuum* L.), are major vegetable crops that have achieved tremendous popularity over the last century. All these vegetables belong to the family Solanaceae. They stand 1st, 4th and 7th according to production of all vegetable crops respectively. They are grown in almost every country of the world — in the field, greenhouses and net houses. These vegetable crops are very versatile and are grown either for fresh market or processing. These vegetables production and consumption has grown quite rapidly over the past 25 years.

Tomato by its nature is a perennial plant, but is commercially cultivated as an annual crop. Greenhouse tomatoes are generally indeterminate and require trellising. Cultivars used in commercial fresh market production under field conditions are determinate and are much shorter in stature than the indeterminate types. Determinate types are easier to harvest and have more concentrated fruit maturity (Rhodes, 2002). At present, tomato is grown in an area of around 3.9 million-hectares worldwide with an annual production of 108 million metric tonnes in the year 2004 (FAO Statistical Database, 2005). Tomato is rich in vitamins A and C and fibre, and is also cholesterol free (Hobson and Davies, 1971). An average sized tomato (148 g) boasts only 35 calories. Tomato contains approximately 20–50mg of lycopene/100 g of fruit weight (Kalloo, 1991). Lycopene is part of the family of pigments known as carotenoids which are natural compounds that create colours of fruits and vegetables. Lycopene is the most powerful antioxidant in the carotenoid family and it protects humans from free radicals that degrade many parts of the body; lycopene is also known to prevent cancer (Rao and Agarwal, 2000). At present, tomatoes are consumed at a higher rate in the developed countries than in the developing countries and hence it may be referred to as a luxury crop.

Eggplant, also known as aubergine, brinjal, or Guinea squash, is in the 4th rank of vegetable crops (FAO, 2005). It has considerable economic importance in Asia, Africa and Subtropics (India, Central America), but it is also grown in some warm temperate regions (Mediterranean area, South of the USA) (Sihachakr *et al.*, 1993). Eggplant was cultivated in 1.7 million ha in the world for a total production of 27. 6 million t, of which 92. 4% of the world productions were covered by Asia (FAO, 2005). Although lower than that of tomato, eggplant nutritious value is comparable to other common vegetables (Grubben, 1977). Its fresh weight is composed of 92.7% moisture, 1.4% protein, 1.3% fibre, 0.3% fat, 0.3% minerals, and the remaining 4% consists of various carbohydrates and vitamins (A and C) (Khan,1979).

Peppers are cultivated all over the world as spice and vegetable crop. Pungent peppers are hot peppers (chillies) that are used for spicing culinary preparations and non-pungent peppers are sweet peppers (paprikas) that are used as vegetables. Pericarp of both hot and sweet peppers turn red at maturity due to carotenoid pigments (orange and brown peppers are also available). Oleoresin, a red oily mixture of carotenoids and other non-volatile compounds is prepared from the dried fruit (pericarp and placenta) and is used in flavouring, improving texture and imparting pleasant colour to foods. Since the prohibition of the use of synthetic food colouring agents, interest in *Capsicum* based oleoresin has been increased which is a natural colouring agent (Galmarini, 1997). Nutritionally peppers are among the richest known plant sources of vitamin C (ascorbic acid), E, B1 (Thiamine), B2 (riboflavin), B3 (niacin) and carotenes (alpha, beta and gamma). Ascorbic acid and carotenes are antioxidants that scavenge free radicals (Colditz, 1987). The literature shows that *Capsicum* is used as a therapeutic agent for cancer (Bosland, 1993) and a cure for rheumatic troubles (Anon, 1964). Green chillies are rich source of a substance called 'rutin' which is of immense pharmaceutical use (Purseglove, 1977) and 'Capsaicin' an alkaloid which is used as a pungent food additive in formulated foods (Johnson *et al.*, 1998). Peppers are grown on a small part of the cultivated area of the world approximately 1 million hectares. Tropical and sub-tropical developing countries account for 80% of the total cultivated area. About 60% of the total production is from Asia, 30% from Europe and 10% each from Africa and America. India is world largest chilli producer.

All these three major vegetable crops suffer from many diseases that are caused by bacteria, fungi, viruses and nematodes: bacterial diseases-bacterial wilt (*Pseudomonas solanacearum*), bacterial canker (*Corynebacterium michiganense*) and bacterial speck (*Pseudomonas syringae*); fungal diseases-fusarium wilt (*Fusarium oxysporum* f. sp. *lycopersici*), verticillium wilts (*Verticillium dahliae*) and early and late blights (*Alternaria solani* and *Phytophthora infestans*), respectively. The nematode *Melidogyne incognita* causes a root-knot disease (Sihachakr *et al.*, 1994). These diseases also contribute to the reduction of hybrid seed recovery from the field. The use of disease resistant cultivars might be an effective way of controlling the above diseases.

To produce disease resistance in these crops through convetional breeding methods take longer time. In order to shorten the duration, biotechnological approaches were used in tomato, eggplant and pepper. Genetic engineering techniques may play a major role in the development of disease resistant cultivars. Plant tissue culture has progressed immensely from its inception in the 1930s, when the scientists used this technique to grow cells in culture. Currently, it is used for many different purposes such as callus induction, anther culture, suspension culture, protoplast culture and somatic embryogenesis and also transgenic prodution. Plant tissue/cell culture is a key facilitator component in genetic transformations using *Agrobacterium tumefaciens*, electroporation and particle bombardment. *In vitro* regeneration of these vegetable crops has been a subject of research because of the commercial value of the crop and its amenability for further improvement via genetic manipulation techniques (Ochoa-Alejo and Ramiraz-Malagon, 2001; Collonnier *et al.*, 2001; Bhatia *et al*, 2004).

Plant Regeneration

Tomato

The development of a cost effective and efficient protocol for mass propagation of high quality seedlings via tomato tissue culture could help lower the price per seedling. A good *in vitro* plant regeneration system may also assist in further improvement of the commercially important cultivars for transferring agronomically important traits via genetic engineering technique (Plate Ia–d; Plate IIa–b). Mass propagation of tomato has been attempted through the use of various methods, including shoot tip culture (Fari *et al.*, 1992; Izadpanah and Khosh Khui, 1992), somatic embryogenesis (Chen and Adachi, 1994; Gill *et al.*, 1995; Kaparakis and Alderson, 2002) direct organogenesis from intact explants (Ichimura and Oda, 1998) or protoplast culture (Muhlbach, 1980; Sink *et al.*, 1986). In tomato, adventitious shoot regeneration can be achieved either directly (Dwivedi *et al.*, 1990) or indirectly through an intermediate callus phase (Geetha *et al.*, 1998). Depending upon the cultivar and the combination and concentrations of growth regulatos, both callus and shoots may be produced together (Bhatia, 2003).

The primary mode of regeneration in tomato is via shoot organogenesis from callus that originates following dedifferentiation of leaf, hypocotyl or cotyledon explants, or directly from a thin layer of cells of pedicel and peduncle (Compton and Veilleux, 1991). In these vegetable crops, the majority of tissue culture attempts have been made to obtain organogenesis rather than somatic embryogenesis or shoot meristem culture. The success in regeneration response has been found to be dependent largely on the genotype, explant and plant growth regulators used in the culture medium (El-Farash *et al.*, 1993).

Most genotypes of tomato respond uniquely to plant growth regulators (PGRs) during regeneration (Kurtz and Lineberger, 1983). Variations in quantity and type of PGRs influence both the percentage of explants responding, and the number of shoots produced by an explant was reported (Plastira and Perdikaris, 1997). These differences are heritable and may be governed by both cytoplasmic and nuclear genes, as illustrated in the reciprocal hybrids developed by Ohki *et al.* (1978). Genotypic differences can be seen for the requirements of PGRs and the type of explant. Frankenberger *et al.* (1981a, b) showed genotypic influences on regeneration. Davis *et al.* (1994) reported that the cultivar 'Better Boy' regenerated only from hypocotyl, whereas 'Spring Giant' regenerated from both hypocotyl and cotyledonary explants. Genotypic differences are also evident within the wild relatives of tomato. In *L. hirsutum* not all genotypes show high regeneration capacity. Shoot morphogenic response in *L. hirsutum* extends from exceptional with numerous shoots for some genotypes to recalcitrant with no shoots being produced for others. Protoplasts of highly responsive genotypes also regenerate exceptionally well (Stommel and Sinden, 1991). A significant general combining ability (GCA) exists for the regeneration ability of the parental lines. This implies that crosses between high regenerating lines result in progenies with high shoot-forming capacities (Frankenberger *et al.*, 1981a, b). Specific PGR requirements for each genotype make the tissue culture task in tomato quite difficult and researchers need to deal with each genotype individually.

Researchers have used various types of explants viz., cotyledon, hypocotyl, pedicel, peduncle, leaf, stem sections and inflorescence for organogenesis. In tomato, the reproductive organs can also be regenerated into vegetative organs. Compton and Veilleux (1991) regenerated *de novo* shoots, roots and flowers using tomato inflorescence explants. Further, Applewhite *et al.* (1994) obtained explants from pedicels and peduncles of flowering tomato plants, which regenerated into roots and shoots and eventually entire fruit bearing plants. The type of explants used not only determines the proportion of explants, which show organogenesis, but also the number of shoots produced per explant. Duzyaman *et al.* (1994) found that the degree of shoot regeneration was in the order of leaves 'cotyledons' hypocotyls, and all cultivars responded similarly. Plastira and Perdikaris (1997) reported differential regeneration frequency of various explants in the order of hypocotyl>cotyledon>leaf. In contrast to these findings, Schutze and Wieczorrek (1987) reported *in vitro* shoot production from cotyledon explants was better than that from hypocotyl explants. Most tissues of tomato seem to have high totipotency; however the choice of the right explant may vary with the genotype (Upender *et al.*, unpublished data).

For tomato regeneration, a wide variety of plant growth regulators have been used at varying concentrations. The concentration of growth regulators employed is dependent on the cultivar being cultured and the particular cytokinin or auxin being employed. Changing the concentration of growth regulator results in variations in the time required for organogenesis and number of shoots formed per explant (Chen *et al.*, 1999; Costa *et al.*, 2000a, b; Venkatachalam *et al.*, 2000; Bhatia *et al.* 2004). Usually plantlets are regenerated either directly (Dwivedi *et al.*, 1990), or from primary callus (Jawahar *et al.*, 1997). Subculture of unorganised callus to a medium in which the ratio of cytokinin to auxin is increased, or in which there is only cytokinin present, leads to shoot differentiation (Gresshoff and Doy, 1972). Four major cytokinins viz. zeatin, 2-iP, BA, and kinetin, can be used either separately or in combination with auxins for organogenesis in tomato. Long-term callus culture was achieved on a medium containing low concentrations of NAA 1.0 mg/l (Imanishi *et al.*, 1976). Vnuchkova (1977a, b) examined the organogenesis on 150 different types of media and concluded that combinations of kinetin and IAA (6. 0 mg/l IAA + 5.0mg/l kinetin or 5.0mg/l IAA + 8.0 mg/l kinetin) are the most suitable for meristem formation in tomato explants. Gunay and Rao (1980) found that a combination of IAA-BA was superior to IAA-kinetin for shoot regeneration. BA or zeatin alone induced shoot formation from leaf callus (Kartha *et al.*, 1976). Zeatin and BA were also found superior to kinetin for shoot formation from tomato leaf explants (Dhruva *et al.*, 1978). Tomato cv. craigella stem explants cultured on an auxin-free medium containing cytokinin produced adventitious shoots at the top of the explant. However, addition of the auxin-transport inhibitor TIBA stimulated caulogenesis with loss of polarity (Cassells, 1979). Pulse treatment with cytokinin was not beneficial for shoot production. No shoots were formed when leaf discs were exposed to high BA concentrations of 10 or 100mM for a short period of 1–6 h (Villiers *et al.*, 1993). Regeneration of shoots from explants was increased by pre-treating source plants with an aqueous solution of the anti-GA_3 compound CCC (DeLanghe and DeBruijne, 1976). The direct inclusion of CCC in the medium was not beneficial. The beneficial effects of CCC on tomato regeneration were corroborated by Carlson *et al.* (1979), with the additional observations that pretreatment with CCC resulted in the highest regeneration frequency.

Several studies have been conducted to understand the role of ethylene in tomato regeneration. Coleman *et al.* (1980) found that neither ethylene nor ethephon stimulated rooting in tomato leaf discs but both inhibited the induction of root formation by IAA. Furthermore, his studies with other auxins and an ethylene inhibitor demonstrated that ethylene was not a rooting hormone for tomato *per se*, as exogenous ethylene did not stimulate rooting. However, Mensuali-Sodi *et al.* (1995) reported that root formation in cotyledonary explant occurred in hormone free LS medium, which was evolving a low quantity of ethylene. Indole acetic acid (20–50μM) increased the ethylene production and inhibited root formation in tomato. Contrary to these findings, Bose *et al.* (1977) demonstrated a significant stimulation of rooting in cuttings treated with ethylene, ethrel or acetylene. For root induction, tomato does not seem to require any exogenous PGR (Mensuali-Sodi *et al.*, 1995). Many researchers have speculated that tomato has a high level of endogenous auxin, based on the observations of shoot cultures producing roots without the addition of auxins in the medium. Unconventional non-purine cytokinins were evaluated by Izvorska *et al.* (1982) on callus growth, and plant or flower bud formation. Greatest effects were achieved with 4-allyl- and phenyl-thioureidosalicylic acids and *N*-allyl- and phenyl-*N*-2-pyridylthiourea. Various not so common PGR precursors have also been studied. Applewhite *et al.* (1994) successfully used isatin for induction of callus, vegetative and floral shoots. Isatin slowly converts into an active auxin, however it rarely induces roots. Branca *et al.* (1990) studied the effects of synthetic auxins, BOA and BIA. For induction and formation of shoots BOA was particularly effective but had a weak activity on cell elongation. In contrast, BIA was more effective in inducing cell elongation and was less active in morphogenesis. Nguyen *et al.* (1992) studied the steroid glycosides for their PGR-like properties on tomato tissue culture, and found that they failed to produce enough PGR-like stimulation. Tomatoside and capsicoside gave some stimulation, which was well below those obtained with the growth regulators. Optimum PGR for tomato is genotypic dependent; however pulse treatment of PGR in general is not found to be beneficial for rooting. Tomato contains high levels of endogenous phytohormones and thus it does not require higher concentrations of auxins for rooting (Mensuali-Sodi *et al.*, 1995).

Eggplant

Fari *et al.* (1991) developed shoot buds on cut hypocotyls of seedlings raised on hormone-free MS medium. From these, 3–30 rooted plantlets/seedlings were produced. Plant regeneration in eggplant is shown in Plate III. Polisetty *et al.* (1994) described a protocol for micro-propagation from leaf explants taken from plants grown in the field or in sterile conditions. The culture was initiated using 1 mg/l IAA+ 2.9 mg/l Kin + 0.1 mg/l BA. Low sugar concentration was important for shoot regeneration. It took 3–4 months for initiation in pots *ex vitro* and 99% of plants survived during the establishment in pots. Ionescu (1998) reported that the regeneration was dependent on explant-type and genotype by using different types of seedling explants of 4 genotypes of *S. melongena*. The explants were cultured on basic MS medium supplemented with auxins (IAA, NAA, IBA) and cytokinins (Kin, BA) in various combinations and concentrations. Differences were noted between genotypes in their regenerative ability, ranging from 74.2% in H3 X G52 to 94.1% in 202-S1. Frequency of induction of shoots (15.1 to 35.4%), callus (60.8–96.7%), somatic embryos (0–27.1) and roots (5–29.0%) differed depending on type of explant and auxin/cytokinin ratios. Direct adventitious budding was reported from hypocotyl tissues.

Magioli *et al.* (1998) described efficient method or organogenesis from calli of eggplant by application of thidiazuron. Leaves and cotyledons were found to be the most responsive explants. Optimal rates of shoot bud induction (75–100 buds/explant) were achieved in the presence of 0. 2 µM TDZ. Organogenic calli were transferred to growth regulator-free MS medium before shoot excision. Rooting was induced on half-strength MS medium supplemented with 0.6 µM IAA. Yamamoto and Matsumoto (1998) obtained differentiated adventitious buds on the hypocotyl-derived callus of eggplant cv. Taya-nasu, but they did not develop into shoots upon subculturing because of recallusing. After, being grafted onto the hypocotyl of a Taya-nasu seedling, the adventitious bud grew into a shoot concurrently, vigorous roots developed on the hypocotyl. These morphogenic responses resulted in plantlets which could be transplanted into soil.

Zelcer *et al* (1983) observed direct shoot regeneration from root without callus formation and subcultured on MS medium without growth regulators. One of the main objectives of micropropagation of eggplant is to produce disease-free planting material or for the elimination of pathogens. In pathological studies maintaining diseased material *in vitro* would be of value. By micropropagation of plant tissues infected with mycoplasma- like organisms (MLOs) is possible to multiply the diseased plants through tissue culture methods. Bhansali and Ramawat (1993) used little leaf-diseased eggplant tissues infected with MLOs to establish a micropropagation system for MLOs within host tissues under controlled conditions. Axillary buds were initially cultured on woody plant medium (WPM) at 25±2°C under continuous light (3000 lux). Sprouted buds proliferated quickly on modified WPM medium containing 1 mg/l BA within 20 days. Diseased shoots produced 7–8 shoots per explant with regular subculturing onto fresh medium. Many shoots (15) were produced on a medium containing 1 mg/l BA and 0. 1 mg/l IAA. Developed shoots produced callus within 20 days and roots within 40 days from the base of the shoots. In this way 40–50 shoots showing little leaf symptoms were obtained in a single 100 ml Erlenmeyer flask. The micropropaged plantlets continuously gave typical symptoms in almost all cultured shoots up to 25 subcultures. Besides visual symptoms, MLOs in micropropagated plantlets were confirmed using mycoplasma-specific DNA stain under light microscopy. Electron micrographs further verified spherical and pleuromorphic MLOs (without cell walls) in the vascular bundles of cultured shoots aged 6 months and 1 year. It is concluded that this finding has wide applications in plant mycoplasmal research for developing immunodiagnostic procedures and *in vitro* screening of antimycoplasmal chemicals. In another investigation, Bhansali and Jindal (1997) used phytoplasma-infected tissues of eggplant to establish a micropropagation system for rapid multiplication of the pathogen within host tissues.

Dormant and sprouted axillary buds, along with nodal segments were successfully cultured on various culture media such as WPM, MS and B5 at 28 ± 2°C under controlled light condition (16 h/8 h cycles of light and darkness). Newly developed shoot buds quickly proliferated in a medium containing 1 mg/l BA. Large numbers of shoots (4–10 shoots/explant) were recorded on elongation medium containing 0.5 mg/l Kin and 0.1 mg/l IAA or IBA. Developed shoots also produced callus tissues at the cut end and at the base of the explant near agar surface. Examination of the micropropagated plants by light and electron microscopy also confirmed that the plants had phytoplasmal pathogens actively multiplying in the host tissues.

Pepper

It is now well established that *in vitro* responses such as adventitious shoot induction, shoot elongation and plant regeneration are genotype-dependent. In order to develop an efficient plant regeneration protocol, a number of commercial varieties were compared for shoot regeneration potential, shoot elongation and rooting of regenerated shoots (Christopher and Rajam, 1994; 1996; Venkataiah and Subhash, 2001, Venkataiah *et al.*, 2003). As a rule, differences due to genotype were as important as or even larger than even caused by media selected as best according to the indications available. Another conclusion calls our attention to the difficulties of repeating experiments with same genotype and method in a different laboratory (Fari and Andrasfalvy, 1994; Steinitz *et al.*, 1999; Ochoa-Alejo and Ramirez-Malagon, 2001). Genotypic differences for tissue culture responses may be related to variation in endogenous hormone levels (Steinitz *et al.*, 1999). Different explants from a single genotype do not respond identically in culture, most likely due to varying gradients of endogenous hormones (Christopher *et al.*, 1991). Even in a single experiment with similar explants culture response is never 100% in most experiments. Many of the genetic differences could be circumvented by growing the source plants under optimal conditions and also by varying nutrients and hormonal composition in the culture medium. Regeneration in tissue culture is a genetically controlled trait in several plants including *Capsicum* (Venkataiah *et al.*, 2001; 2003). Gunay and Rao (1978) reported adventitious shoot bud formation and plant regeneration from seedling explants of *C. annuum* cvs California Wonder and Pimento and *C. frutescens* cv Bharath (Plate IVa–e). In hypocotyl segments of *C. annuum* cv Hatvani induction of adventitious shoots and complete plant regeneration reported (Fari and Czako, 1981). The gradual decline in the morphogenetic potential was demonstrated (Fari and Czako, 1981). The correlation between endogenous content of phytohormones and the morphogenetic responses was well documented by Fari *et al.*, (1982), and also they summarised the histological details of shoot bud regeneration (Fari,1983). Shoot proliferation and plant regeneration from mature embryo culture of *C. annuum* cv Mathania was reported by Agrawal and Chandra (1983). Phillips and Hubstenberger (1985) studied the *in vitro* organogenesis of four pepper varieties. Induction of adventitious shoots and plantlet formation has been reported in *C. annuum* cv G_4 from mature zygotic embryo cultures (Christopher *et al.*, 1986). Shoot bud induction and plant regeneration from cotyledon explants of *C. annuum* cv Yutsufusa has been reported (Sripichitt *et al.*, 1987). Successful shoot bud induction and plant regeneration from excised cotyledon explants of *C. frutescens* has been reported (Subhash and Christopher, 1988). Among various species and genotypes of *Capsicum* tested, *C. baccatum* var. *baccatum* and *C. baccatum* var. *pendulam* were regenerated shoots but *C. annuum, C. chinenese* and *C. frutescens* did not show any shoot bud formation after Seedling Decapitation Method (Fari *et al.*, 1995a). Valera-Montero and Ochoa-Alejo (1992) reported successful regeneration in decapitated seedling explants of *C. annuum* cvs Chile de Anqua, Salvatierra and Tampiqueno 74. Ezura *et al.* (1993) observed direct organogenesis without external source of growth regulators in halved seeds of pretreament of water soaking 0 to 5 days, on the cut surface of the hypocotyl (decapitated), in 14 cultivars of *C. annuum.*

In pepper, adventitious shoot bud induction, proliferation and plant regeneration has been reported using diverse explants such as half seed explants (Ezura *et al.*, 1993; Binzel *et al.*, 1996a), zygotic embryos (Christopher *et al.*, 1986), cotyledon, hypocotyl, leaf (Fari and Andrasfalvy, 1994; Steinitz *et al.*, 1999; Ochoa-Alejo and Ramirez-Malagon, 2001), and petiole explants (Christopher, 1990). Among the various explants tested, leaf explants were more pronounced for adventitious shoot bud formation followed by zygotic embryos, cotyledons, hypocotyl and petiole explants (Venkataiah *et al.*, 2003).

In general, addition of 2,4-D to the medium causes callus induction and proliferation in diverse explants. The addition of low levels of cytokinin (Kn or BA) induced high frequency of callus over 2,4-D alone. Although most of the earlier research on cell and tissue culture focussed on the media manipulations, however, the hormonal composition of the medium and other nutrient components have not been as productive in inducing regenerable cultures. Among various cytokinin and auxin combinations tested; BAP in combination with IAA found to be more superior to induce adventitious shoots from different explants. According to Ezura *et al.* (1993), the presence of radicle cells in the mature embryo axis explant was critical for organogenesis of adventitious buds. However, Valera-Montero and Ochoa-Alejo (1992) suggested that roots do not influence bud formation in hypocotyl tissues of chilli pepper but rather have a stimulatory effect on the bud elongation, which is an important step in shoot production. The process of adventitious regeneration in pepper cotyledon explants has been widely documented in the literature (Fari and Andrasfalvy, 1994; Steinitz *et al.*, 1999; Ochoa-Alejo and Ramirez-Malagon, 2001; Venkataiah *et al.*, 2003).

Somatic Embryogenesis

Micropropagation through somatic embryogenesis provides an efficient method for producing large of number of elite or transgenic plants. The embryo induction process results from a delicate balance of cytokinin and auxin ratio in plant tissue. In addition to PGRs, other compounds/factors that have induced embryogenesis include heavy metals, proline, salicylic acid and medium pH. etc.

Tomato

In tomato, somatic embryos have been induced from protoplast cultures (Chen and Adachi, 1994), hypocotyls (Newman *et al.*, 1996) and from intact seedlings growing on a medium containing high concentrations of cytokinin (Kaparakis and Alderson, 2002). However, efficient procedures for large-scale propagation of tomato through somatic embryogenesis are not yet available despite recognition of the importance and usefulness of somatic embryogenesis and synthetic seeds for micropropagation of F1 hybrids (Gill *et al.*, 1995). Somatic embryogenesis in tomato begins in subepidermal cortical tissue. Developing somatic embryos are encased in a distinct protoderm, and have a suspensor-like structure that is loosely attached to the hypocotyl tissue. Newman *et al.* (1996) and Gill *et al.* (1995) reported that somatic embryogenesis occurs directly (without an intervening callus proliferation phase) from the meristematic centres formed in the subepidermal region of the nodular tissues developed at the basal ends of induced seedlings raised in *in vitro* culture. Regenerated somatic embryos thus developed can be successfully grown into whole plants. Cytokinins appear to play an important role in somatic embryogenesis in tomato (Praveen *et al.*, unpublished data). Gill *et al.* (1995) developed a simple one-step method for somatic embryogenesis. Tomato seeds were germinated on MS medium containing 50 or 80 µM BA, and the resultant seedlings were maintained for 6 weeks on the same medium. Somatic embryos were developed on the lower hypocotyl region. Kaparakis and Alderson (2002) used this protocol to successfully reproduce the results, not only in tomato but also in aubergine (egg plant) and pepper. Somatic embryogenesis in tomato is still at infancy, and substantial research is still required to be undertaken to obtain high quality somatic embryos, which can be used directly or may be converted into artificial seeds for propagation of high value hybrid cultivars.

Eggplant

Different types of explants like shoot tip, hypocotyl, leaf and stem were used to initiate callus induction in eggplant. Yamada *et al.* (1967) reported callus induction and purification from zygotic embryos from two different cultivars of eggplant. Hypocotyl sections of eggplant did not respond to 2,4-D (Kamat and Rao, 1978), however NOA or NAA showed good response to callus initiation (Matsuka and Hinata, 1979). Ammirato (1983) reported that plant regeneration from eggplant tissue culture may be obtained via organogenesis. He was able to demonstrate that auxin was critical for induction of somatic embryos. Production of numerous somatic embryos from callus derived from leaf explants was recorded when cultured on MS medium supplemented with 54 µM NAA (Gleddie *et al.*, 1982). While comparing several media, MS salts were found to be superior to four others including B5 salts. It was observed that when the level of NH_4NO_3 supplied in B5 medium was increased to the level present in MS, the frequency of embryogenesis was elevated to the same extent as that of MS. Sucrose at the level of 0.06 M in the medium was found to be sufficient to produce embryogenesis in optimal frequency.

Fobert and Webb (1988) described the effects of polyamine precursors and polyamine biosynthetic inhibitors on somatic embryogenesis in eggplant. Cotyledons grown on MS medium with NAA formed callus, roots and somatic embryos. Low levels of NAA (0.1–0.5 mg/l) induced rhizogenesis, intermediate levels (1.0–5.0 mg/l) promoted embryogenesis and high levels (10. 0–50. 0 mg/l) favoured callogenesis. Addition of polyamines or their precursors did not induce morphogenesis on medium containing no growth regulators, nor did it affect embryogenesis on medium containing NAA except at the highest concentration tested, which was inhibitory. Enzyme activated inhibitors of putrascine synthesis significantly reduced embryogenesis and stimulated rhizogenesis. Saito and Nishimura (1994) described an improved culture condition for somatic embryogenesis using an aseptic ventilative filter in eggplant. Non-vitrified somatic embryos were formed which developed to the cotyledonary stage on transfer to hormone-free MS medium. A novel group of genes were described by Momiyama *et al.* (1995) whose expression significantly increased the changes in eggplant morphology during early stages of somatic embryogenesis. The nucleotide sequences of 8 cDNA fragments were determined and homologues compared with those stored in a DNA database. The genotype and explant play an important role in both organogenesis and somatic embryogenesis. Sharma and Rajam (1995a) extensively studied the influence of these factors on somatic embryogenesis. Hypocotyl, cotyledon and leaf explants were tried with combination of BA and NAA. NAA was found to induce embryogenesis in all the three explants but hypocotyl required higher concentrations of the hormone in comparison to the other two explants. Among the explants, hypocotyl yielded the maximum number of adventitious shoots followed by cotyledon and leaves. Interestingly, significant differences in morphogenetic potential were reported within a single explant. According to them there was a basipetal gradient for organogenesis (i.e. increase in the number of shoots from base to apex), while the terminal hypocotyl segments showed better embryogeneic potential than the median segment (Sharma and Rajam, 1995b).

Afele *et al.*, (1996) analyzed the molecular basis of differential cultivar response to the induction of somatic embryos by using digoxigenin in eggplant cvs. Wase Shinkuro (WS) and Kumamoto Naga (KN). WS produced embryogenic callus which was capable of differentiated into mature somatic embryos. While KN cultures, treated similarly, failed to differentiate into somatic embryos. Digoxigenin-differential display of 0,1,2,4 and 10-day-old cultures of both cultivars revealed several PCR products derived from WS but not from KN at a crucial stage during embryogenesis. These results indicated that cultivar differences in embryogenic callus induction may be attributed in part to differences in mRNA expression during the culture process. In eggplant, explants from different regions of the leaf showed significant differences for embryogenic potential. Discs from the apical region of leaves yielded more somatic embryos than those from the basal region (Yadav and Rajam, 1997). Putrescine treatment promoted somatic embryogenesis and at 0.5 mM it caused about a 6-fold increase in the number of somatic embryos, accompanied by an increased putrescine titre. On the other hand, the authors reported that spermidine and spermine had no stimulatory effect on embryogenesis, rather they were inhibitory at higher concentrations. All tested inhibitors of polyamine biosynthesis significantly inhibited somatic embryogenesis. These results demonstrated that: (i) a positive correlation between the spatial distribution of free and conjugated polyamines and the embryogenic capacity of an explant, and (ii) stimulation of somatic embryogenesis by using putrescine, suggesting an intricate regulatory role of polyamines, specifically putrescine in somatic embryogenesis of eggplant.

Four critical stages of embryogenesis from leaf discs of eggplant were identified by scanning electron microscopy (Yadav and Rajam, 1998). Temporal changes in arginine decarboxylase (ADC) activity and polyamines (PAs) were observed during critical stages of embryogenesis. High levels of PAs, especially putrescine, due to higher ADC activity in discs from the apical region (with high embryogenic capacity) than from the basal region of leaf (with poor embryogenic capacity) were correlated with differential embryogenic response. Kinetic studies of the up and down regulation of embryogenesis revealed that putrescine and difluoromethylarginine pretreatments were most effective before the onset of embryogenesis.

Highly efficient somatic embryogenesis in eggplant was achieved from suspension culture by elevating the concentration of gelrite in the regulation medium in combination with an aseptic ventilation filter as a capping material in the culture vessel (Saito, 1999).

Pepper

Somatic embryogenesis and plant regeneration has been reported in *C. annuum* using immature zygotic embryos. Harini and Lakshmi Sita (1993) first described direct somatic embryogenesis from immature zygotic embryos in chilli pepper. Somatic embryos were also obtained through calli i.e. indirect somatic embryogenesis (Buyukalaca and Mavituna, 1996). The production of artificial seeds, consisting of somatic embryos encapsulated in calcium alginate gel beads has been reported in *C. annuum* (Buyukalaca *et al.*, 1995). Direct somatic embryogenesis was observed in cotyledon and leaf explants of *C. baccatum*, on the media supplemented with various concentrations of 2,4-D in combination with 0.5 mg/l. 2.0 mg/l Kn produced maximum number of somatic embryos from cotyledon and leaf explants (Venkataiah *et al.*, unpublished data). In *C. annuum*, somatic embryogenesis was observed either in the presence of 2,4-D + TDZ or BAP or 2,4-D and Coconut milk, but not with 2,4-D alone (Harini and Lakshmi Sita, 1993; Binzel *et al.*, 1996; Kintzios *et al.*, 2001). In many cvs, it is evident that cytokinins can exert promotive effect on somatic embryogenesis. A cytokinin is required for maturation and germination of somatic embryos in *C. annuum* (Binzel *et al.*, 1996).

Haploidy

Tomato

Changes were observed in the gametes during the culture of anthers with pollens at different stages of development (meiosis, tetrads, uninucleate pollen and mature pollen) and best stages for inducing embryoid formation were those involving uninucleate or binucleate and mature pollen (Lee *et al.*, 1999). Genotypic differences also play a role in androgenic haploid production in tomato (Lee *et al.*, 1999). Haploid callus cultures induced from anthers were successfully obtained in selected tomato races. Callus differentiated could be controlled both in the dark and in the light by adding hormones (NAA and Kin at various levels) to defined minimal media. Development into plantlets was achieved only in the light. Haploid shoots can be regenerated from anther callus by simply transferring the callus from high auxin (5. 0mg/l NAA) and low cytokinin (0.01mg/l kin) medium to another medium containing these PGRs in the reverse ratio (0.1mg/l NAA and 2.0mg/l kin).

In vitro cultures of anthers of either self-incompatible diploids and two self-incompatible autotetraploid clones of *L. peruvianum* for callus induction and organ regeneration showed diploid, triploid, tetraploid, hexaploid and octaploid plants and cytochimeras. The occurrence of self-incompatibility in regenerated S-heteroallelic triploids revealed that, besides S-heteroallelic pollen, a balanced hormonal combinations or a certain level of heterozygosity is necessary for the manifestation of self-compatibility (Ramulu and Denereux, 1979). Zamit *et al.*, (1980) cultured anthers from 15 male-sterile mutants and their respective fertile counterparts *in vitro*. The stages at which microsporogenesis was disrupted in the mutants ranged from before meiosis to the microspore stage. Only anthers of mutant with ms 1035, in which mircosporogenesis was arrested at the tetrad stage, consistently produced callus. The frequency of callus induction in anthers of this mutant was highest, 37–46% in anthers cultured shortly after the intiation of the beakdown of meiosis, when anthers were 2.0–2.5 mm in length. Callus induction occurred in less than 19% of cultures of fertile anthers. Callus was derived in part from sporogenous tissue. Cells of the tissue ranged in chromosome number from x to 12x.

Isolated uninucleate microspores of cv. HS102 were cultured on DG medium supplemented with NAA and Kin or modified MS medium (Varghese and Yadav, 1986). Callus formation was enhanced by cold treatment of anthers before cultures and more calli formed on agar than on liquid medium (Chlyah and Taarji, 1984). The addition of charcoal to the medium was beneficial. Longer anthers developed more vigorously than shorter anthers (Santana, 1985). Anthers collected in the field gave a higher degree of organogenesis than those from greenhouse-raised plants. Pretreatment at low temperature gave good results. Best results in Roma msms (homozygous for the recessive gene ms1035) were recorded with a 48 h pretreatment at 4°C (92.6% of anthers producing shoots), while in Roma Msms (heterozygous for the gene) 9 days pretreatment at 10°C (4.5% of anthers producing shoots) proved most effective. MS medium supplemented with 2ip was better than Zeatin. Best results from irradiation were obtained with a dose of 400 rad alone or in combination with cold pretreatment. Most of the regenerants obtained were mixoploid, but haploids, diploids, triploids and tetraploids were also found (Zagorska *et al.*, 1982). Oankh *et al.* (1986) cultured anthers of cultivar Roma and its male-sterile counterpart carrying ms 1035 and developed haploid, diploid and mixoploid plants. The regenerants (477) included plants resembling the source (42% of male-sterile and 34% of normal plants) and also differing appreciably from it. Plants of the former group were mainly mixoploid. After mixoploids, haploids and near haploids were the second most frequent types. Zamorzaeva and Smirnov (1987) cultured explants at the premeiotic stage; meiosis could subsequently be studied in the microsporocytes *in vitro*. Explants taken at the earliest stages of microsporogenesis needed a more complex nutrient medium than those taken later. Even gelling agents affect the androgenesis. Anthers from the cultivars L680A, Ailsa Craig and Licato were plated on DBM1 medium solidified with bacto-agar, gelrite, noble agar or phytagar to evaluate their ability to promote initiation and growth of anther callus. Both the number and diameter of calli were affected by the type of solidifying agent and genotype of the anthers. Significant interactions were also found between cultivars and solidifying agent. Bacto-agar reduced the number and size of callus by 38% when compared with the liquid treatment and by 42% when compared with the best agar treatment (noble agar) (Jaramillo and Summers, 1990).

Shcehreva *et al.*, (1990) isolated anthers at various stages from 2 forms (homozygous and heterozygous for the male sterility gene ms 1035) form each of 8 varieties and cultured on MS medium modified in various ways (MS1 and MS2 with different growth regulators for rooting and regeneration). The best results were obtained by using anthers 2–2.5 mm in size taken from homozygous form for the recessive *ms* gene, the greatest morphogenetic potential was observed in cvs Roma (msms), Vigapol (msms) and Pearson (Msms). There were differences in response to the cytokinins added to the nutrient medium; on MS1 with 2iP, cvs Roma (msms) and Vigapol (msms) gave the best morphogenetic response but on MS2 with zeatin the best were POR (msms), cvs Roma (msms) and Sar (Msms). Zagorska *et al.*, (1998) tested the androgenetic ability of 85 tomato genotypes. Callus was induced from anthers of 53 lines and hybrids. Regeneration of plants was obtained only from calli of 15 genotypes. The *in vitro* response of anthers from the cultivars Roma, Pearson, San Marzano, Por, Sar, Vigapol, Day, David and Start, containing the *ms1035* gene, confirmed the dependence of expression on both callus induction and organogenetic potential based on the homozygous or heterozygous nature of gene.

Jaramillo and Summers (1991) used three tomato cultivars to determine the effect of nine dark/light regimes for callus initiation and growth from anthers. Calli mass and diameter was increased as the dark period increased. Callus diameter peaked at 8 weeks of dark treatment in cvs Ailsa Craig and Licato, where as in cv L-680A calli continued to grow over the entire 10-weeks dark incubartion, callus quality and appearance decreased noticeably during this period. Summers *et al.* (1992) determined the microspore developmental stage and anther length on callus production. Anthers of cvs L680A, Licato and Ailsa Craig tomato were plated on Doy's basal medium. Although calli were induced at all stages of anther development, anthers containing prophase I-stage microspores produced the highest frequency of calli. Fewer calli were produced as microspores

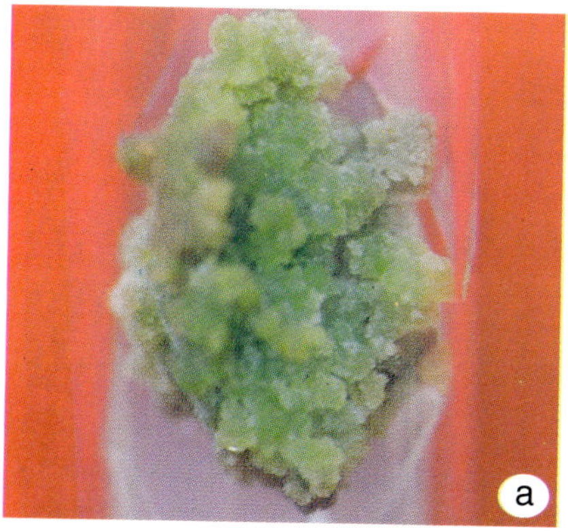

Green and regenerative callus induction on MS medium supplemented with 2.0 mg/l 2,4-D and 1.0 mg/l Kin from cotyledon explant.

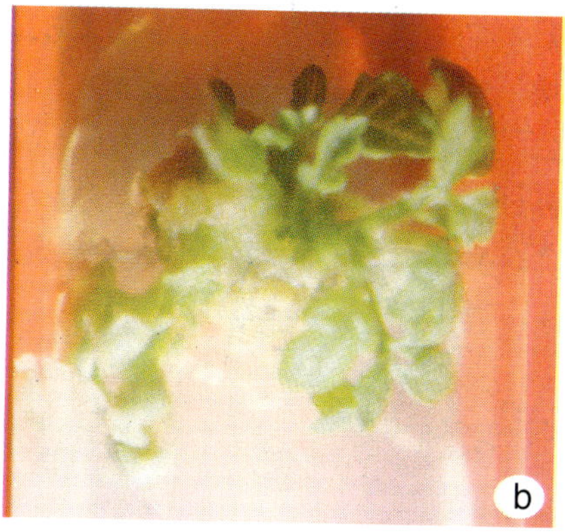

Multiple shoot induction on MS medium supplemented with 3.0 mg/l BAP and 0.5 mg/l IAA.

Elongation of multiple shoots on MS medium supplemented with 1.0 mg/l BAP.

Rooting of elongated shoots on MS medium supplemented with 1.0 mg/l IAA.

PLATE I

a–d: Plant regeneration in tomato

Adventitious shoot buds from cut ends of hypocotyl explants after two weeks of incubation (2.0 mgl^{-1} TDZ).

Adventitious shoot buds from cut surface of cotyledon explants after two weeks of incubation (2.0 mgl^{-1} TDZ).

Complete regenerated plantlet on rooting medium (1.0 mgl^{-1} IAA).

Adventitious shoot bud proliferation all over cut surface of leaf explants after two weeks of incubation (2.0 mgl^{-1} TDZ).

Potted regenerated plant of *C. annuum* L. ev X-235 after four weeks of transplantation.

PLATE IV

a–e: *In vitro* plant regeneration in pepper

approached the uninulceate and binucleate pollen stage. Callus diameter was also decreased as the anther development progressed. Significantly larger calli were produced from prophase I than later-stage of anthers. Time of anther and flower bud length were both significantly correlated with anther developmental stage, the number of anthers producing calli and mean calli diameter. In each case, anther length exhibited a significantly better correlation than bud length. The experiments carried out by Ma *et al.* (1999) with Fms-2line demonstrated that the developmental stage of the explant was a crucial factor in callus formation and that optimum stage was between premeiosis and metaphase I, which corresponds to the stage just after the blockage of microsporogenesis. Callus was efficiently induced from anthers, pretreated at 4°C for 3 days. The data obtained clearly showed that the androgenesis was genotype dependent in tomato.

Eggplant

Production of haploid plants through anther culture of eggplant has been reported (Isouard *et al.*, 1979; Vaulx and Chambounet, 1982). These studies suggested that the anthers required pretreatment with high temperature (35°C). A temperature of 25°C produced 3.4 plants per 100 anthers while 8-day pretreatment at 35°C yielded an average of 12 plants per 100 anthers. The auxin in the medium was found to be greatly influenced the ploidy level of the regenerated plants. With 2,4-D, 15–50% regenerants were found to be diploid, while with NAA only haploid plants were regenerated. Callus induction and plantlet formation from freshly isolated microspores of eggplant was reported by Miyaoshi (1996). Culturing of these microspores in a sucrose-free medium was found to be effective for induction of callus. Regeneration of shoots was also successful from these calli in a medium containing zeatin and IAA (Ugandhar *et al.*, unpublished). Still further work is needed for producing the androgenic haploids in eggplant.

Pepper

Genetic expression of major characters in sweet pepper lines derived by anther culture has been studied by many workers. Chen (1985) found in the study of $74H_2$ and $5H_3$ lines of *C. annuum* var. *grossumi* derived by anther cultures that there was considerable variation in the botanical and agronomic characters between lines derived from the same parent. Characters were uniform within lines and between generations of the same line. More than 30% of the lines were superior to the original material induced by anther culture in total yield and 28% had a lower disease index. Some 20%were superior to control Tongfeng 37. The main fruit characters of lines derived by anther culture from a sweet × hot *C. annuum*

F_1 hybrid grown on up to 5 generations (H_1–H_5) inherited from the parent were uniform (Jiang and Li, 1984). A new sweet pepper (*C. frutescens* var. *grossum*) cultivar Haihua 3 was developed by *in vitro* culture from uninucleate pollen of the breeding line 78203. Anther culture was used to produce maintenance lines of the Hungarian variety Feherozon Synthetic (FS), which is composed of 8 different genotypes but identical phenotypes which were stable for the 5 main traits (growth habit, fruit colour, flavour, shape and position) and homozygous (Venczel and Mityko, 1995). Li *et al.*, (1995) developed 6 cultivars directly from anther culture and have been cultivated over 38,400 ha throughout China.

Many reports have shown on anther culture of *Capsicum* in order to produce dihapliods with many desirable traits. Gametoclonal variation in doulbled haploid lines (DH) of pepper was detected by Munyon *et al.* (1989). The lines were evaluated in the field for a number of traits, including plant height, flower morphology, yield per plant and fruit quality. In general, DH lines were shorter, while for the other characters both beneficial and deleterious variations were observed. Particularly noted was a DH line of cv. California Wonder which had an upright fruit orientation due to lack of bending in the pedicel. Genetic analysis indicated that this was due to a single gene mutation. Another DH line produced a few seeds (Morrison, 1987). Anther culture has also been used to develop double haploid lines showing resistance to different diseases and pests. Abak *et al.* (1982) evaluated a total of 27 doubled haploids of *C. annuum* plants from the cross PM217 (resistant) × Yolo Wonder (susceptible) for both root and stem resistance to *Phytophthora capsici.* There was a continuous gradation between complete susceptibility and resistance and a good correlation was observed between the results of both tests. Daubeze *et al.* (1989) reported that the resistance of *C. annuum* line CM334 to strain S197 of *P. capsici* and to pathotype of tobacco mosaic tobamovirus, which remains stable even at temperatures >30°C, was genetically analysed using 2 sets of double haploid lines obtained by anther culture: 30 lines from the F1 hybrid CM334 x Yolo Wonder and 38 lines from the F_1 hybrid VAT × CM334 (VAT being susceptible to the virus but partially resistant to the fungus). The lines were tested for resistance to both pathogens at 22 and 32°C. Results inidicated that the virus resistance controlled by a single gene was also responsible for the stability of resistance at high temperature. Tests indicated that this gene was allelic to L1 and it was named Ll c. It was not genetically linked to the genes for resistance to *P. capsici,* most of which were found effective irrespective of temperature.

Vaulx *et al.* (1982) used a total of 212 doudled haploid lines obtained from six heterozygous (L+/Ll) *C. annuum* hybrids between plants susceptible (L+/L+) and Ll/Ll) to tobacco mosaic virus (Po pathotype) and found that the resistance : susceptibility ratio was 1:1. Selassie *et al.* (1986)

produced 4 homozygous lines, resistant to 3 strains of potato virus Y and one of pepper veinal mottle virus (PVMV) from the F_1 of Perennial x Florida. Daubeze *et al.* (1989) made crosses between the line H_3 (bred in Ethiopia by crossing a local population with a Kenyan cultivar) with resistance to *Levellula taurica* and susceptible Vania. Doudled haploids were obtained from the F_1s by *in vitro* culture and colchicine treatment. When 77 doubled haploid lines were compared with the parents and F_1 under artificial infection, about an eighth of the lines were as resistant as the resistant parent. At least 3 genes appeared to control resistance, which was not associated with poor fruit set. In subsequent studies Shifriss *et al.* (1992) could fit the segregation into simple genetic model. They used the F_1 (HV12), which is immune to *L. taurica* races in Israel, originated from a cross between resistant Ethiopian variety H_3 and susceptible Vania. Following anther culture, haploid plants obtained from the F_1 were diploidized using colchicine and exposed to the pathogen. The F_1 HV12 was among the lines selected. F_1 progeny, obtained by crossing HV12 to susceptible line 1646, was screened in the greenhouse. Seeds from resistant F_2 plants were used to establish 48 F_3 lines in the field; 11 were susceptible, 8 were resistant and 29 segregated for susceptibility and resistance.

Hwang *et al.* (1998) screened 33 breeding lines of pepper for resistance to bacterial spot (*Xanthomonas campestris* pv *vesicatoria*). Plants derived from anther culture of resistant lines were screened again and analyzed for inheritance of resistance. Nine lines including PI 163184 were resistant to disease and PI 183439 and PI 244669 were moderately resistant. Pretreatment of anthers at 35°C for 8 days in darkness was more effective than 4°C treatment for plant regeneration. Anthers at late uninucleate to binucleate stage responded better than those at other stages in inbred line1-1 × PI 163292 and inbred line 5xPI 183481 combinations although in MS line 3 × PI 183431 anthers at early uninucleate stage were more responsive. Large variations in the length of the main stem and floral buds were observed in anther-derived plantlets. Most of the anther-derived plants were moderately resistant to bacterial spot.

Inheritance of resistance to *Meloidogyne* sp. was studied by Hendy *et al* (1985). The line PM217, derived from PI201234 (Central America) and which has resistance to *M. arenaria, M. incognita, M. javanica* and an unidentified *Meloidogyne* species designated Seville, was crossed with the cultivar Yolo Wonder, which has some resistance to *M. javanica,* as was the line PM687, derived from PI322719 (India) and which is resistant to *M. arenari, M. incognita,* and *M. javanica.* Twenty-five doubled haploids from the cross PM217x Yolo Wonder and 32 from the cross PM 687xYolo Wonder were inoculated with 5 populations of the 4 species in pots under controlled conditions in a greenhouse. The results suggested that PM217 has 2 resistance genes, probably independent, designed *Me* 1, controlling resistance to *M. arenaria, M. incognita* and *M. javanica*, and *Me* 2, controlling to *M. javanica* and the unidentified species and that PM687 has 2 genes, probably linked, designated *Me 3*, controlling *M. incognita, M. javanica* and *M. arenaria* (except for one strain) and *Me 4*, controlling resistance to this one strain of *M. arenaria*. It is suggested that *Me1* and *Me 3*, though not allelic, are on the same chromosome. Yolo Wonder had a fifth gene *Me5* controlling its resistance to *M. javanica*. Szasz *et al* (1997) attempted to transfer the regeneration capacity of inbred *C. annuum* line 40017-13 to commercial non-regerenable pepper cultivars (*C. annuum*) by sexual crossing. Two doubled haploid lines derived from the cross 40017R-13 x Tarq3 gave higher regeneration rates than control cotyledon or rooted hypocotyl explants. The obtained hybrids were stabilized by haploid induction.

Protoplast Culture and Somatic Hybridization

Isolated plant protoplasts are cells, which have their cell wall removed by mechanical action or enzymatic digestion. Theoretically all plant cells are totipotent and, therefore, protoplasts may be able to resynthesize the cell wall, divide, form colonies and calli, and, eventually regenerate a new plant. The capacity of plant regeneration from protoplasts will be used as an important tool in plant breeding and physiological studies (Ochoa-Alejo and Ramirez-Malagon, 2001). The development of reproducible and efficient systems of isolation, cultivation and plant regeneration from protoplasts is a fundamental prerequisite for using these techniques.

Tomato

Compared to cultivated tomato, its wild counterparts such as *L. pimpinellifolium, L. peruvianum* and *L. glandulosum* show better regeneration capabilities (Lech *et al.*, 1996). Leaf explants of *L. peruvianum* demonstrated higher morphogenic potential than *L. esculentum*, while the response of another wild relative of tomato *Solanum pennellii* varied with the type of medium used (Tal *et al.*, 1977). Lech *et al.* (1996) found that *L. peruvianum* not only showed better morphogenic potential, but it also responded quickly (2 weeks earlier) compared to *L. esculentum* (Lech *et al.*, 1996). Protoplast cultures of various *Lycopersicon* spp. show similarity in their response to intact explants. Muhlbach (1980) attempted to regenerate protoplasts derived from the leaves of wild *L. peruvianum* and cultivated tomato *L. esculentum*, and found that under the same conditions, *L. peruvianum* regenerated successfully but not the *L. esculentum*. A similar trend was observed by Zapata (1980) with *L. peruvianum* mesophyll protoplasts, which grew and divided actively in liquid medium whereas those obtained from *L. esculentum* did not. Successful plant protoplasts isolation from *L. esculentum* and plant regeneration from the isolated protoplasts in the medium containing BA and 2iP or GA and 2iP has also been described by Shahin (1985). Takashina *et al.* (1998) found new molecular markers that were linked to high shoot regeneration capacity in the wild relative of tomato *L. chilense*. In *L. peruvianum*, a dominant allele at the locus Rg near the middle of chromosome 3 determines the efficiency of shoot regeneration (Koornneef *et al.*, 1993). This finding was also supported by Faria and Illg (1996) who suggested that in *L. pimpinellifolium* a qualitatively inherited dominant trait governs regeneration ability and this can be easily transferred to cultivated tomato. Successful attempts have been made to transfer the superior regeneration capacity of *L. peruvianum* into cultivated tomato through backcrossing (Koornneef *et al.*, 1993). Furthermore, *L. pimpinellifolium* has been used with success as a donor parent to introgress *in vitro* regeneration capacity in recalcitrant tomato cultivars (Faria *et al.*, 2002). Other unconventional breeding procedures such as protoplast fusion have also been attempted to transfer regeneration capability from *L. peruvianum* to *L. esculentum* (Wijbrandi *et al.*, 1988, 1990) with success. Atrazine resistant plants could be recovered from interspecific protoplast fusion of *Solanum nigrum* to *L. esculentum* (Jain *et al.*, 1988). Intergeneric protoplast fusion between *L. esculentum* and *Solanum lycopersicoides*, and shoot regeneration has also been successfully achieved by Sink *et al.* (1986).

Eggplant

In order to overcome the sexual barriers or to improve the fertility of hybrids usually, somatic hybridization experiments via protoplast fusion have been performed to facilitate the introduction of agronomically important traits from wild relatives into cultivated eggplant. These characters were mainly resistant against diseases and parasites, and particularly resistance to bacterial and fungal wilts, nematodes, mites and fruit borers (Sihachakr *et al.*, 1994). Somatic hybridisation provides a means of increasing nuclear and cytoplasmic variability and transferring desirable agronomic traits without detailed genetic or molecular knowledge of the genes encoding for such characters (Jones, 1988). Protoplast fusion requires the control of reliable protocols for protoplast isolation and culture, and plant regeneration in particular. In most cases, mesophyll tissues of eggplant have become the primary choice as a high yield of protoplasts (Bhatt and Fassuliotis, 1981; Jia and Potrykus, 1981). Stems and petioles were also shown to be competent source of protoplasts (Sihachakr and Ducreux, 1987; Sihachakr *et al.*, 1993). A reliable protocol for plant regeneration in eggplant was based on the use of *in vitro* plants as the competent protoplast source, and the high dilution of the protoplast culture on day 15, before transfer of the resulting calli to the regeneration medium supplemented with 2 mg/l zeatin and 0. 1 mg/l IAA (Sihachakr *et al.*, 1993). The first cytological events initiated when protoplasts differentiate to form plantlets have been studied by Fourvier *et al.* (1995), who showed that an increased cytoplasmic and nuclear activity preceeded development of the cell wall and cell proliferation in eggplant. Both chemical and electrical fusion procedures have successfully been used to produce somatic hybrids of eggplant with its wild relatives. The first somatic fusion between eggplant and *S. sisymbrifolium* regenerated 26 aneuploid somatic hybrids. They had only the *S. sisymbrifolium* chloroplast genome, and were highly resistant to root-knot nematodes and potentially resistant to spider mites (Gleddie *et al.*, 1986). However, because of the high hybrid sterility, their usefulness in a breeding program is seriously limited. A total of 10 and 19 tetraploid somatic hybrids of eggplant with *S. torvum* were recovered after chemical and electrical fusion respectively (Guri and Sink, 1988a; Sihachakr *et al.*, 1989a). Most of them had the eggplant cpDNA type, and were all resistant to *Verticillium* wilt, nematodes and partially resistant to spider mites (Guri and Sink, 1988a; Sihachakr *et al.*, 1994). Electrofusion has successfully been used for the production of 83 somatic hybrids of eggplant with *S. khasianum*, another interesting wild relative with resistance to shoot and fruit borer (Sihachakr *et al.*, 1988). Six hybrid plants were examined for the chloroplast genome, and all contained the eggplant cpDNA type. Resistance to the herbicide atrazine has been transferred from the atrazine-resistant biotype *S. nigrum* into somatic hybrids of eggplant by using chemical (Guri and Sink, 1988b) and electrical (Sihachakr *et al.*, 1989b) procedures of protoplast fusion. All the somatic hybrids had *S. nigrum* cpDNA, conferring resistance to 0. 1 M atrazine. Electrofusion has successfully been used for the production of highly fertile somatic hybrids of eggplant with *S. aethiopicum* (Daunay *et al.*, 1993). Field evaluation revealed that the somatic hybrids showed a better pollen fertility (30–85%), compared to their sexual counterparts with only 20–50%. Moreover, some hybrid clones yielded a very good fruit production up to 9 kg/plant, which was 3 to 4 times as high as that of the parental lines, while the sexual F1 hybrids produced only very few fruits. Moreover, field evaluation carried out in Bogor (Indonesia) revealed that the somatic hybrids tested were resistant against bacterial wilt (*Ralstonia solanacearum*). Other fertile somatic hybrids between eggplant and *S. integrifolium* (now considered as *S. aethiopicum*), have been obtained and characterized morphologically and molecularly (Rotino *et al.*, 1995, 1998). Most of them, as well as some of their self and backcrossed progenies, were resistant to *Fusarium* wilt.

Lastly, intergeneric crosses, generally unfeasible by sexual hybridisation, have been performed in eggplant via protoplast fusion. A selection scheme has been developed to recover two intergeneric somatic hybrids of egg-plant with a triple mutant, chlorophyll deficient, streptomycin and kanamycin-resistant *Nicotiana tabacum* as a fusion partner (Toki *et al.*, 1990). Somatic hybrids of eggplant with a transformed sexual hybrid between tomato and *Lycopersicon pennellii* were also produced by somatic hybridisation. Only hybrid calli with leaf primordia were first recovered (Guri *et al.*, 1991), then normal plants were regenerated after irradiating one of the fusion partners with Gamma rays (Liu *et al.*, 1995; Samoylov *et al.*, 1996; Samoylov and Sink, 1996). Although, the combination of complete genomes can easily be obtained, the resulting somatic hybrids, particularly between remote species, were often sterile, thus limiting their use in eggplant breeding program. Because of their amphiploid status, intensive backcrosses are needed for the introgression of desirable traits into the cultivated eggplant. Moreover, when the fusion partners are distantly related, the final product is often an asymmetric combination of the two genomes (Pijnacker *et al.*, 1989; Gleba *et al.*, 1984; Pelletier *et al.*, 1985). Therefore, asymmetric fusion could be an alternative promising approach to produce fertile asymmetric hybrids bearing few chromosome fragments from one parent associated with a complete chromosome set from the other parent (Jones, 1988; Sihachakr *et al.*, 1994). Indeed, fertile hybrids with a tolerance to *Verticillium* wilt, and particularly a morphology close to the cultivated eggplant, were recovered after asymmetric fusion between eggplant protoplasts and x-rays irradiated protoplasts of *S. torvum*, a distant eggplant relative (Jarl *et al.*, 1999). The usefulness of the tetraploid symmetric somatic hybrids in breeding programs depends on their ability to be backcrossed to the recurrent genotype of eggplant. Their ploidy level must be reduced to a dihaploid status by haploidisation methods before backcrossing.

Pepper

One of the most important breeding characteristics of the *Capsicum* is its resistance to diseases, especially those by virus and bacteria (Peter *et al* 1993). There are several wild species of the genus, which have been recognized as sources of resistance. Interspecific incompatibility and/or hybrid sterility, however, has hindered the transfer of desirable characteristics by conventional hybridization methods. Thus genetic engineering methods are an important tool in solving this problem (Galmarini, 1997). The isolation and regeneration of plants from protoplasts and *in vitro* flowering reported in *Capsicum annuum* cv california wonder by Saxena *et al.* (1981). Diaz *et al.* (1988) also reported the regeneration of plants from leaf protoplasts of *Capsicum annuum* cvs Americano and Dulce Italiano. Plant regeneration from mesophyll protoplasts of *C. annuum* cv california wonder has been demonstrated (Prakash *et al.*, 1997).

Genetic Transformation

A genetic engineering program requires: delivery of appropriate DNA into the cell;— integration of the introduced DNA into the chromosome for stable transformation;— selection of transformed cells (promoters and markers); and a good *in vitro* regeneration system essential for an effective genetic engineering system that seeks to exploit genetically transformed plants for commercial applications.

Tomato

Several primitive species of the genus *Lycopersicon*, especially *L. peruvianum*, *L. hirsutum* and *L. glandulosum*, represent an important source of genes, conferring resistance against various diseases and pests of cultivated tomatoes (Lukyanenko, 1991). However, introduction of those genes into commercial cultivars of *L. esculentum* by conventional breeding techniques often encounters serious difficulties due to high incompatibility barriers (Kaul, 1991). To overcome these problems certain modern approaches of gene manipulation might be required. Genetic engineering techniques are being used in tomato variety improvement programmes to fight these challenges. Antisense RNA technology in tomato, which has been used to increase the shelf-life (Flavr Savr tomato) was released in 1994 by Calgene, Inc. Flavr Savr was the first to receive FDA approval (Stone, 1994). Several genes involved in fruit development and ripening have been characterised. One of these genes, encoding pectin methylesterase, was used to genetically engineer tomato fruit. An antisense gene for pectin methylesterase in tomato has been introduced to create high solid fruits (Janick, 1993). Expression of an antisense RNA gene in genetically engineered fruit reduced the enzymatic activity of pectin methylesterase up to 95%. Salt resistant tomato which can grow up to 200mM NaCl (up to 40% of sea water salinity) has been developed by the modification of a trait that encodes a protein that exports excess salt to vacuole before it damages the tissues. Modified plants can flourish in salty water, and can also produce edible fruits. Although the leaves accumulated high sodium concentrations, the tomato fruit displayed very low sodium content (Zhang and Blumwald, 2001). To enhance the carotenoid content and profile of tomato fruit, transgenic lines were produced by Romer *et al.* (2000) containing a bacterial carotenoid gene (crtl) encoding the enzyme phytoene desaturase, which converts phytoene into lycopene. Expression of this gene in transgenic tomatoes did not elevate total carotenoid levels. However, the beta-carotene content increased about three-fold, up to 45% of the total carotenoid content. The current protocols used for tomato transformation are based on shoot regeneration from leaf disc/cotyledon tissue co-cultivated with disarmed

Agrobacterium tumefaciens harbouring binary vectors (Fillatti *et al.*, 1987; Bird *et al.*, 1988). The efficiency of such procedures is generally low (Hamza and Chapeau, 1993; Frary and Earle, 1996) because most of the transformed leaf/cotyledon cells could not develop into shoots (Peres *et al.*, 2001).

Eggplant

Genetic engineering provides the eggplant breeders with new tools to complement and supplement sexual hybridisation for improvement of existing varieties or creation of totally new germplasm, by insertion of genes encoding for agronomically useful traits. Eggplant is well adapted to the application of these techniques. It responds well to *Agrobacterium*-mediated transformation with both cointegrate and binary vectors carrying *nptII*, *CAT*, and *GUS* genes (Guri and Sink, 1988c; Rotino and Gleddie, 1990; Leone *et al.*, 1993; Fari *et al.*, 1995b; Iannacone *et al.*, 1995). Furthermore, regeneration efficiency from transformed explants of eggplant can be enhanced by using growth regulators, such as thidiazuron (TDZ – 0. 1 μM), and antibiotics, like augmentin (300 μg/ml) and kanamycin (50 μg/ml) (Billings *et al.*, 1997). Although, several useful genes conferring resistance against bacterial and fungal wilts as well as insects are now available (Zhu *et al.*, 1994; Kombrink and Sonsich, 1995; Schuler *et al.*, 1998; Kumar *et al.*, 1998a), genetic transformation of eggplant and its related species using genes of interest is still in its early stages. So far, two fields have successfully been explored: genetic manipulation for resistance against insects using *Bacillus thuringiensis* endotoxin genes (Rotino *et al.*, 1992b; Arpaia *et al.*, 1997; Kumar *et al.*, 1998a, b), and parthenocarpic fruit production (Rotino *et al.*, 1997). Eggplant has successfully been transformed via the *Agrobacterium*-mediated transformation method, using binary vectors harbouring only marker genes, like *nptII* gene (Guri and Sink, 1988c; Filippone and Lurquin, 1989; Rotino and Gleddie, 1990). Mendelian inheritance of the functional *nptII* gene was demonstrated through three selfed generations of transformed eggplant (Sunseri *et al.*, 1993). In order to develop eggplant resistance to Colorado Potato Beetle (*CPB*) (*Leptinotarsa decemlineata* Say), one of the major pests in Europe and North America, attempts were first made to transfer wildtype genes of *Bacillus thuringiensis* var. Tolworthi (Bt) encoding for a coleopteran insect-specific toxin via *Agrobacterium tumefaciens* into eggplant, but their expression was too low for successful insect control (Rotino *et al.*, 1992b; Chen *et al.*, 1995). A mutagenized Bt gene (*CryIIIB*) with an increased transgene expression was developed and successfully introduced into eggplant (Arpaia *et al.*, 1997). The presence of the Cry3B toxin in leaf extracts was confirmed in 57 out of 93 transgenic plants tested by DAS-ELISA assay. Insect bioassay revealed significant insecticidal activity on neonate larvae of CPB in 23 among 44 plants tested, and the progeny from self-pollination showed the same level of resistance as the parental lines. Similar results were obtained by successful transfer in eggplant of synthetic *Cry1Ab* (Kumar *et al.*, 1998b) and *CryIIIA* (Jelenkovic *et al.*, 1998) genes, conferring resistance to shoot and fruit borer (*Leucinodes orbonalis*) and CPB respectively. Field trials showed that Bt *CryIIIB* expressing lines were able to control heavy CPB natural infestations and could be exploited to develop an integrated pest management strategy for eggplant cultivation (Acciarri *et al.*, 2000). Another interesting approach in genetic engineering of eggplant is the development of parthenocarpic fruits (Rotino *et al.*, 1997), which are an attractive feature that can increase consumer acceptance. In fact, auxin levels were manipulated during fruit development after successful introduction of a chimeric gene containing the coding region of the *iaaM* gene from *Pseudomonas syringae* pv. *savastanoi* into eggplant under the control of the ovule-specific *DefH9* promoter from *Antirrhinum majus*. In the absence of pollination, the transgenic eggplants developed seedless parthenocarpic fruits without any supply of exogenous phytohormones, even at low temperatures, which normally prohibit fruit production in untransformed lines. Fruits obtained from emasculated flowers of transgenics had the same size and weight as those derived from pollinated flowers of the untransformed controls (Rotino *et al.*, 1997). When pollinated, the parthenocarpic plants produced seed full fruits indistinguishable in colour, consistency, weight and size, from the berries obtained by hand-pollination of the controls. The lines carrying the *DefH9-iaaM* transgene could also be used for the production of seedless marketable fruits in open field cultivation by incorporating into the genome of the hybrid seed a dominant gene conferring male sterility (Rotino *et al.*, 1997). Trials in an unheated greenhouse showed that transgenic parthenocarpic eggplant hybrids gave a significantly higher production than untransformed controls and a commercial hybrid. Moreover, it is noteworthy that the treatment of floral buds with exogenous plant growth regulators did not affect the production of transgenics, whereas the production of all the untransformed hybrids was positively influenced by growth regulator sprays (Donzella *et al.*, 2000). Recently, resistance to the root-knot nematode *M. incognita* has been engineered by inserting the *Mi-1* gene from tomato. The four transformed eggplant lines stably expressed the resistance trait through generations (Frijters *et al.*, 2000).

Pepper

Liu *et al.* (1990) worked on *Agrobacterium*-based *in vitro* reneration and transformation system in bell pepper using 6 cultivars and one wild Guatemalan accession. The wild accession exhibited the best regeneration response. Only occasional elongation of shoot buds was achieved in Yolo Wonder Local by culture in the dark on a medium containing 10 mg/l BA and 1 mg/l IAA. Transformed shoot buds and leaf-like structures were obtained, showing beta-glucaronidase activity predominantly in the vascular and perivascular tissues with no indication of contamination with *Agrobaterium* in the tissues. Attempts to regenerate whole transgenic plants from transformed shoot buds were unsuccessful. Wang *et al.* (1991) reported the regeneration and transformation of sweet pepper (*C. frutescens*). Segments of cotyledon, hypocotyl and leaf taken from 10 to 30 day-old seedlings of 3 cultivars Shuangfeng, Zhongjiao 2 and Zhongjiao 3 were cultured on MS medium supplemented with various phytohormones for induction of callus, shoots and roots. *Agrobacterium tumefaciens* transformed plantlets were obtained using strains C58 and GV3111 and Horch's leaf disc method. The shooting frequency from transformed explants was 100% on MS medium supplemented with 4–6 mg/l BA + 0.5 mg/l IAA; 35% of these shoots elongated on MS medium containing 2 mg/l zeatin or 2 mg/l BA + 1–3 mg/l GA_3. Shoots were rooted on MS medium supplemented with 0.1-0.5 mg/l NAA. Regenerants were transplanted into soil with a 100% survival rate and developed into normal plants. Transient expression of the *GUS* gene was detected and kanamycin-resistant shoots were obtained from cotyledons.

Zhu *et al.* (1996) obtained transgenic sweet pepper plants from *Agrobaterium*-mediated transformation. Fertile transgenic sweet pepper (*C. annuum* var. *grossum*) plants were regenerated at high rates from explants that were cocultivated with *A. tumefaciens* strain GV311-SE harbouring a plasmid that contains the cucumber mosaic cucumovirus coat protein (*cms-cp*) gene. Young leaves were most effective for bud induction and subsequent shoot elongation. Whereas whole hypocotyls were the most ineffective. Southern blotting analysis of DNA isolated from putative transgenic plants revealed that•3 out of 5 R_1 plant lines contained the *cmv-cp-* gene. Western blot analysis of CMV-CP containing R_1 plants showed that two of them accumulated significant levels of the foreign gene product whilst the other two expressed it only at low levels. Transformation experiments carried out on rooted hypocotyl and cotyledon explants of the cv. 40017R-13 using *A. tumefaciens* were largely unsuccessful. Many adventitious buds were induced on infected explants, but they were lost during selection by kanamycin. Rooted shoots were not obtained on kanamycin-containing selection medium. Antibiotics interacted with the growth regulators (BAP and TDZ) inhibiting shoot regeneration.

Siregar and Sudarsono (1997) obtained shoot regeneration from hypocotyl segments of hot pepper mediated by non-disarmed isolates of *Agrobacterium.* Hypocotyls isolated from 2-week-old seedlings of hot pepper cvs Hot Spiral, Jatibala, Lari and Tit Super were cultured on growth medium which was either uninoculated or inoculated with *A. tumefaciens* or *A. rhizogenes* and their growth was observed for at least 3 months. Hypocotyls cultured on MSO without *A. tumefaciens* or *A. rhizogenes* did not develop any further during the 3 months in culture. However, explants cultured on inoculated growth medium formed callus from the infection site and from other parts of the hypocotyl in contact with the pathogen between 5 and 14 days after inoculation. While only 32 and 37% of Jatibala hypocotyl formed callus, 65 and 74% of Laris hypocotyls formed callus after inoculation with *A. tumefaciens* or *A. rhizogene* respectively. Inoculation with *A. rhizogenes* resulted in some root formation from Jatibala, Laris and Tit Super hypocotyls between 30 and 36 days after inoculation. Both *A. tumefaciens* and *A. rhizogenes* were also able to induce direct shoot meristem regeneration in a small proportion of cases in these 3 cultivars. However, initiated shoot meristems usually developed into single or double leaves that did not develop any further. When hypocotyls with shoot initials were transferred onto MS medium containing 2 mg/l GA_3 shoot elongation occurred.

Manoharan *et al.* (1998) established a protocol for regeneration and *Agrobacterium*-medaited genetic transformation in hot chilli (*C. annuum* cv. Pusa Jwala). High frequency regeneration of shoot buds from cotyledonary leaves was achieved using MS medium supplemented with 0.5 mg/l TDZ. Elongation of shoot buds and subsequent rooting was obtained on half-strengh MS medium with 0.5 mg/l IAA. The *A. tumefaciens* strain EHA 105, carrying a binary vector plasmid pBI 121, was used for the transformation. The T-DNA contained the *nptII* gene, driven by the nopaline synthase promoter and terminator sequences, which provides resistance to kanamycin and was used as selectable marker. The beta-glucuronidase (gus) gene, driven by the CaMV 35S promoter, was used as a reporter gene. The cotyledonary explants from *in vitro* grown seedlings were cultivated for 2 days. Shoot buds were produced on a regeneration medium containing 50 mg/l kanamycin. The transgenic nature of regenerated plants was confirmed by histochemical staining of GUS, PCR and Southern hybridization anlyses of the nptII gene. Lim *et al.* (1999) attemped regeneration and genetic transformation of hot pepper plants. High frequency of plant regeneration was observed when hot pepper cotyledon segments were cultured on MS medium supplemented with 0. 1 mg/l NAA and 2.0 mg/l zeatin or 10 mg/l IBA. An addition of 5 μM $AgNO_3$ to these media improved the regeneration frequency by up to 8%. Verification of presummed transgenic plants that survived in selection media was done by amplification of selectable marker genes by PCR. Successful expression of the ADA (adenosine deaminase) gene was detected by Northern blot and enzyme activity, suggesting a potential application of ADA as a selectable marker in genetic transformation. The *bar* (herbicide resistance) gene was also introduced into hot pepper and progenies of the transgenic plants were obtained. Stable inheritance (3:1 ratio) of *bar* was confirmed by PCR analysis and treatments with of 5 mg/l basta.

Dong *et al.* (1992, 1995) obtained transgenic pepper plants containing a CMV (cucumber mosaic cucumovirus) satelite RNA cDNA by *A. tumefaciens*-mediated transforamtion. Transformants were regenerated on selection medium and some flowered and set fruit. PAGE and Northern blot analyses demonstrated the expression of CMV sat-RNA gene in the transgenic plants. Compared with non-transgenic plants, most transgenic plants had 10 days delay in disease symptoms, lower disease index and slow CMV development. Kim *et al.* (1997) investigated satellite-RNA-mediated resistance to cucumber mosaic virus in progeny of transgenic plants of hot pepper cv. Golden Tower that express CMV satellite RNA. The transgenic plants were self-fertilized and their progenies were tested for stable inheritance and expression of the cDNA of CMV satellite RNA. PCR and RNA gel blot analyses showed that the introduced gene was stably transmitted and expressed in the progeny. Symptom attenuation in the offspring was confirmed upon inoculation with CMV-Y or CMV-Korea (CMV-Kor) strains.

CMV-cp (cucumber mosaic-cucumovirus coat protein) gene transformation into pepper and expression in the offsprings of the transgenic plants has been reported by Zhang *et al.* (1994). *CMV-cp* genes were introduced into pepper by coculturing cotyledon explants of 3 cultivars with *A. tumefaciens.* Transformants were selected on the basis of kanamycin resistance and regenerated at a transformation rate of 10%. Southern and Northern blotting and ELISA indicated that the *CMV-cp* gene was incorporated, transcribed and translated in transgenic plants and their offsprings. Inoculation with the virus showed that the transgenic plants and their offsprings were resistant to CMV. Genetic transformation of hot pepper with ribosome inactivating proteins (RIPs), which are known to have cytotoxic on eukaryotic cells by cleaving a specific adenine residue of 28S rRNA, has been carried out to improve fungal disease resistance (Kim *et al.*, 1995). A reconstructed vector (pGA-RIP) was transformed into hot pepper using the *A. tumefaciens* system. Fourteen regenerants were confirmed as transgenic and all were fertile.

Application of modern genetic manipulation has been limited in pepper (*C. annuum* L.) due to the lack of an efficient transformation system. Following the development of an efficient protocol for *in vitro* regeneration of pepper cotyledons, we investigated the key factors affecting transformation and established a highly efficient genetic transformation system using the pepper cotyledon as starting material. In this system, cotyledon explants are preconditioned for 2 days on kanamycin (km)-free DM1 (differentiation medium) [Murashige and Skoog (MS) salts/Gamborg

B5 vitamins basal medium supplemented with 20 g/l sucrose, 5,000 mg/l DJ nutrients and a hormone combination of 1.0 mg/l indoleacetic acid (IAA) and 5. 0 mg/l 6-benzyladenine (BA) solidified with 0. 7% agar, pH 5. 8], followed by co-cultivation with *Agrobacterium tumefaciens* on DM1 for 2 days and delay selection on DM1 with 500 mg/l carbenicillin (carb) for 2 days. The explants are then placed on DM1 containing 10 mg/l $AgNO_3$, 50 mg/l km-sulfate and 500 mg/l carb. After 4–5 weeks, the explants with buds were transferred to EM1 (elongation medium) (MS salts/Gamborg B5 vitamins basal medium supplemented with 20 g/l sucrose, 5,000 mg/l DJ nutrients, 10 mg/l $AgNO_3$ and a hormone combination of 1.0 mg/l IAA, 3.0 mg/l BA and 2.0 mg/l gibberellic acid, solidified with 0.7% agar, pH 5.8) with 50 mg/l Km and 500 mg/l carb for the elongation of buds. After 3–6 weeks, 3- to 4-cm-long elongated shoots were excised and planted on RM1 medium (MS basal medium supplemented with a hormone combination of 0.2 mg/l NAA and 0. 1 mg/l IAA, solidified with 0. 8% agar, pH 5.8) with 25 mg/l km and 200 mg/l carb for rooting. Li *et al.* (2003) tested four genotypes of pepper, and all presented a high differentiation efficiency (81.3% on average), elongation rate (61.5%) and rooting efficiency (89.5%). Polymerase chain reaction analysis results showed that 40.8% of the regenerated plantlets were transgenic plants (Li *et al.*, 2003).

Lee *et al.* (2004) used two genes, TMV-CP and PPI1 (pepper-PMMV interaction 1 transcription factor), to transform commercially important chilli pepper (*C. annuum*) inbred lines (P915, P409) by means of *Agrobacterium* co-culture. Eighteen independently transformed T0 plants were obtained. The most critical point in the pepper transformation protocol was the selection of shoots growing on calli — referred to as callus-mediated shoot formation (indirect shooting) — because shoots not grown from the callus (direct shooting from the wounded surface) developed into non-transformants. Selection of the correct right callus type also proved to be an important requirement for obtaining transformed peppers. Six different types of calli developed during the selection process. Shoots regenerated from two of these types, while one type regenerated significantly more shoots than the other types, suggesting that the capacity for shoot formation is callus type-specific. Although the transformation rate was low, transformation via callus-mediated shoot formation proved to be reproducible and was confirmed by Southern and Northern blot analyses. Based on the experimental data, Lee *et al.* (2004) have succeeded in developing a new protocol for the selection and transformation of pepper and this protocol can be used in the future for pepper transformation.

Conclusions and Prospects

The evaluation of genetic resources in these vegetables, so far mainly based upon phenotype, revealed numerous useful traits in its wild relatives, but very less data of molecular markers are available for their characterisation. Efforts are being made on their development through the techniques of RFLP, RAPD and AFLP. These studies should allow for, first, the enrichment of the existing genetic linkage map and development of new ones, and, second, an acceleration of the identification and isolation of markers and genes involved in resistance to pests and pathogens, useful for MAS and further gene transfers. The ability of these vegetable crops to respond well in tissue culture, particularly plant regeneration from cultured seedling explants (mainly leaf, cotyledon and hypocotyl), cells and protoplasts, has allowed the application of various and powerful biotechnological techniques for management of genetic resources and improvement of these crop plants. Production of haploids and somaclonal variants has been reported. Genetic engineering studies in these vegetable crops need to be further exploited for the introduction of new genes encoding resistance to biotic and abiotic stresses (Arpaia *et al.*, 1997; Jelenkovic *et al.*, 1998; Kumar *et al.*, 1998a). Recently, transgenic plants were shown to offer several advantages, like low cost, flexibility and rapidity, for production of recombinant proteins (Artsaenko *et al.*, 1998; Zeitlin *et al.*, 1998; Vaquero *et al.*, 1999), such as single-chain Fv (scFv) antibodies, which have strong medical and pharmaceutical interest (Bird *et al.*, 1998). Similarly, approaches to improve the nutritive value in vegetable fruit based on genetic engineering might significantly contribute to generate materials designed for a better supply of vitamins and essential amino acids for human nutrition particularly for the third world population. Nevertheless, special attention has to be paid to the use of selectable marker genes conferring resistance in plant cells against antibiotics or herbicides, which provide a reliable system to effortlessly select rare transformed plant cells after co-introducing the desired gene along with the selectable marker. Beyond the laboratory, these markers have no role and their presence in crops and food has provoked much public concern. A new selection system, involving genes for enzymes, phospho-mannose isomerase (Joersbo *et al.*, 1998) or xylose isomerase (Haldrup *et al.*, 1998), which metabolise mannose-6-phosphate and xylose respectively, may be a better alternative to the use of antibiotic marker genes. In eggplant, as in many crop plants, rapid progress is being made in biotechnology. A close interaction and collaboration between biotechnologists, plant breeders, pathologists and agronomists will be essential to transfer the laboratory findings to the fields. Though much work is in progress for the improvement of these vegetable crops, using genetic engineering techniques, there is a need to develop marker free transgenic plants.

REFERENCES

Abak, K., E. Pochard and R. Dumas De Vaulx. 1982. Transmission of resistance to *Phytophthora capsici* on roots and stems of pepper plants. Study of DH. lines issued from cross PM 217 x Yolo wonder through anther culture. *Capsicum Newsletter*, **1:**62–64.

Acciarri, N., G. Vitelli, S. Arpaia, G. Mennella, F. Sunseri, and G.L. Rotino. 2000. Field protection against Colorado Potato Beetle of Bt-expressing transgenic eggplant. *Hort. Sci.*, **35:**722–725.

Afele, J.C., Y. Tabei, T. Yamada, T. Momiyama, F. Takaiwa, T. Kayano, S. Nishimura and T. Nishio. 1996. JARQ, **30:**175–179.

Agrawal, S., and N. Chandra. 1983. Differentiation of multiple shoot buds and plantlets in cultured embryos of *Capsicum annuum* L. cv Mathania. *Curr Sci.*, **52:**645–646.

Ammirato, P.V. 1983. Embryogenesis, *In*: Hanbook of plant cell culture, D.A. Evans, W.R. Sharp, P.V. Ammirato and Y. Yamada (eds.), Macmillan Pub Co., New York, Vol. 1, pp. 82–123.

Anon. 1964. The vegetable for treating rheumatic pains. New development in *Capsicum* growing. *Ital Agric.*, **101:**187–188.

Applewhite, P.B., R. Kaur-Sawhney and A.W. Galston 1994. Isatin as an auxin source favoring floral and vegetative shoot regeneration from calli produced by thin layer explants of tomato pedicel. *Plant Growth Reg.*, **15:**17–21

Arpaia, S., G. Mennella, V. Onofaro, E. Perri, F. Sunseri and G.L. Rotino. 1997. Production of transgenic eggplant (*Solanum melongena* L.) resistant to Colorado potato beetle (*Leptinotarsa decemlineata* Say). *Theor. Appl. Genet.*, **95:**329–334.

Artsaenko, O., B. Kettig, U. Fiedler, U. Conrad and K. Düring. 1998. Potato tubers as a biofactory for recombinant antibodies. *Mol. Breed.*, **4:**313–319

Bhansali, R. and K.G. Ramawat. 1993. Micropropagation of little leaf diseased egg plants infected with mycoplasma like organisms. *J. Hort. Sci.*, **68:**25–30.

Bhansali, R.R. and S.K. Jindal. 1997. *In*: Recent Advances in Management of Ecosystem, pp. 391–396.

Bhatia, P. 2003. Optimisation of physical, chemical and biological factors for *in vitro* micropropagation through direct regeneration, axillary branching and somatic embryogenesis methods for the 'Red Coat' cultivar of tomato (*Lycopersicon esculentum* Mill.). Doctoral thesis. Primary industries research centre, Central Queensland University, Rockhampton, Australia.

Bhatia, P., N. Ashwath, T. Senaratna and D. Midmore. 2004. Tissue culture studies of tomato (*Lycopersicon esculentum*). *Plant Cell Tiss. Org. Cult.*, **78:**1–21,

Bhatt, D.P. and G. Fassuliotis. 1981. Plant regeneration from mesophyll protoplasts of eggplant. Z. Pflanzenphysiol., **104:**481–489.

Billings, S., G. Jelenkovic, C.-K. Chin and J. Eberhadt. 1997. The effect of growth regulation and antibiotics on eggplant transformation. *J. Amer. Soc. Hort. Sci.*, **122:**158–162.

Binzel, M.L., N. Sankhla, S. Joshi and D. Sankhla. 1996. Induction of direct somatic embryogenesis and plant regeneration in pepper (*Capsicum annuum* L.). *Plant Cell Rep.*, **15:**536–540.

Bird, R.E., K.D. Hardman, J.W. Jacobsen, S. Jonhson, B.M. Kaufman, S.M. Lee, T. Lee, S.H. Pope, G.S. Riordan and M. Whitlow. 1988. Single chain antigen binding proteins. *Science*, **242:**423–426.

Bose, T.K., T.P. Mukherjee and R.N. Basu. 1977. Effects of ethylene and acetylene on the regeneration of adventitious roots. *Indian J. Plant Physiol.*, **20:**134–139.

Bosland, P.W. 1993. Breeding for quality *Capsicums*. *Capsicum and Eggplant Newsletter*, **12:**25–31.

Branca, C., A. Torelli and M. Bassi 1990. Effects of benzisoxazole and enzisothiazole on tomato plant regeneration *in vitro*. *Plant Cell Tiss. Org. Cult.*, **21:**17–19.

Buyukalaca, S. and F. Mavituna. 1996. Somatic embryogenesis and plant regeneration of pepper in liquid media. *Plant Cell Tiss. Org. Cult.*, **46**:227–235.

Buyukalaca, S., F. Mavituna and M.L. Gomez-Gullamon. 1995. Artificial seeds of pepper somatic embryos. *Acta. Hort.*, **412**:106–110.

Carlson, C., L. Conviser and J. Preece. 1979. Programming stock plants for tissue culture success. *In*: 6th Annual Meeting Plant Regulator Work Group, Las Vegas Nevada, pp. 197–204.

Cassells, A. 1979. The effect of 2,3,5-triiodobenzoic acid on calogenesis in callus cultures of tomato and pelargonium. *Physiol. Plant*, **37**:239–246

Chen, H., J. Zhang, T. Zhuang and G. Zhou. 1999. Studies on optimum hormone levels for tomato plant regeneration from hypocotyls explants cultured *in vitro*. *Acta Agric. Shanghai*, **15**:26–29.

Chen, L.Z. and T. Adachi. 1994. Plant regeneration via somatic embryogenesis from cotyledon protoplasts of tomato (*Lycopersicon esculentum* Mill). *Breed. Sci.*, **44**:337–338.

Chen, Q., G. Jelenkovic, C. Chin, S. Billings, J. Eberhardt and J.C. Goffreda. 1995. Transfer and transcriptional expression of coleopteran *CryIIIB* endotoxin gene of *Bacillus thuringiensis* in eggplant. *J. Am. Soc. Hort. Sci.*, **120**:921–927.

Chen, X.S. 1985. Determination of combining ability and analysis of heterosis in pollen lines of *Capsicum annuum* var. *grossum* Sendt. *Acta Hort. Sinica*, **12**:267–272.

Chlyah, A. and H. Taarji. 1984. Androgenesis in tomato. Plant tissue and cell culture application to crop improvement. Czechoslovak Ac, pp. 241–242.

Christopher, T. 1990. Mutation breeding and tissue culture studies in *Capsicium*. Ph.D. Thesis, Kakatiya University, Warangal, India.

Christopher, T. and M.V. Rajam. 1994 *In vitro* clonal propagation of *Capsicum* spp. *Plant Cell Tiss. Org. Cult.*, **38**:25–29.

Christopher, T. and M.V. Rajam. 1996. Effect of genotype, explant and medium on *in vitro* regeneration of red pepper. *Plant Cell Tissue Organ Cult.*, **46**:245–250.

Christopher, T., B. Prolaram and K. Subhash. 1991. Differential *in vitro* morphogenetic response in hypocotyl segments of *Capsicum annuum*. *Indian J. Exptl. Biol.*, **29**:68–69.

Christopher, T., B. Prolaram, M.V. Rajam and K. Subhash. 1986. Plantlet formation in embryo culture of *Capsicum annuum* L. var G4. *Curr Sci.*, **55**:1036–1037.

Colditz, G.A. 1987. Beta-carotene and Cancer. *In*: Horticulture and Human Health, Contributions of fruits and vegetables. ASHS symposium series no. 1. B. Quebedeaux and FA. Bliss (eds.), Prentice-Hall, NJ. pp. 150–159.

Coleman, W.K., T.J. Huxter, D.M. Reid and T.A. Thorpe. 1980. Ethylene as an endogenous inhibitor of root regeneration in tomato leaf discs cultured *in vitro*. *Physiol. Plant*, **48**:519–525.

Collonnier, C., I. Fock, V. Kashyap, G.L. Rotino, M.C. Daunay, Y. Lian, I.K. Mariska, M.V. Rajam, A. Servaesl, G. Ducreux and D. Sihachakr. 2001. Applications of biotechnology in eggplant. *Plant Cell Tiss. Org. Cult.*, **65**:91–107.

Compton, M.E. and R.E. Veilleux. 1991. Shoot, root and flower morphogenesis on tomato inflorescence explants. *Plant Cell Tiss. Org. Cult.*, **24**:223–231.

Costa, G.M., F.T.S. Nogueira, W.C. Otoni and S.H. Brommonschenkel. 2000a. *In vitro* regeneration of processing tomato (*Lycopersicon esculentum* Mill.) 'IPA-5' and 'IPA-6'. Ciencia e Agrotecnologia, **24**:671–678.

Costa, M.G.C, F.T.S. Nogueira, M.L. Figueira, W.C. Otoni, S.H. Brommonschenkel and P.R. Cecon. 2000b. Influence of the antibiotic timentin on plant regeneration of tomato (*Lycopersicon esculentum* Mill.) cultivars. *Plant Cell Rep.*, **19**:327–332

Daubeze, A.M., A. Palloix and E. Pochard. 1989. EUCARPIA. VII meeting on genetics and breeding on Capsicum and Eggplant, Kragujevac, Yugoslavia, June 27–30, pp. 229–232.

Daunay, M.C., M.H. Chaput, D. Sihachakr, M. Allot, F. Vedel and G. Ducreux 1993. Production and characterization of fertile somatic hybrids of eggplant (*Solanum melongena* L.) with *Solanum aethiopicum* L. *Theor. Appl. Genet.*, **85**:841–850.

Davis, D.G., K.A. Breiland, D.S. Frear and G.A. Secor. 1994. Callus initiation and regeneration of tomato (*Lycopersicon esculentum*) cultivars with different sensitivities to metribuzin. *Plant Growth Reg. Soc. Am. Quart.*, **22**:65–73.

DeLanghe, F. and E. DeBruijne. 1976. Continuous propagation of tomato plants by means of callus culture. *Science Hort.*, **4**:221–227.

Dhruva, B., T. Ramakrishnan and C.S. Vaidyanathan. 1978. Regeneration of hybrid tomato plants from leaf callus. *Curr. Science*, **47**:458–460.

Diaz, I., R. Moreno and J.B. Power. 1988. Plant regeneration from protoplasts of *Capsicum annuum*. *Plant Cell Rep.*, **7**:210–212.

Dong, C.Z., C.X. Jiang, L.X. Feng and J.Z. Guo. 1992. Transgenic pepper plants (*Capsicum annuum* L.) containing CMV Sat-RNA cDNA. *Acta Hort. Sinica*, **19**:182–184.

Dong, C.Z., C.X. Jiang, L.X. Feng, S.D. Li, Z.H. Gao, J.Z. Guo and D.W. Zhu. 1995. Transgenic tomato and pepper plants containing CMV Sat-RNA. *Acta Hort.*, **402**:78–86.

Donzella, G., A. Spena and G.L. Rotino. 2000. Transgenic parthenocarpic eggplants: superior for increased winter production. *Molecular Breeding*, **6**:79–86.

Dumas De Vaulx R., D. Chambonnet and E. Pochard. 1981. Culture *in vitro* d' anthers de pigment (*Capsicum annuum*) amelioration de taux de' obtention des plantes chez differents a + 33oC. Agronomie, **1**:859–860.

Duzyaman, E., A. Tanrisever and G. Gunver. 1994. Comparative studies on regeneration of different tissues of tomato *in vitro*. *Acta Horti*, pp. 235–242.

Dwivedi, K., P. Srivastava, H.N. Verma and H.C. Chaturvedi. 1990. Direct regeneration of shoots from leaf segments of tomato (*Lycopersicon esculentum*) cultured *in vitro* and production of plants. *Indian J. Exp. Biol.*, **28**:32–35.

El-Farash, E.M., H.I. Abdalla, A.S. Taghian and M.H. Ahmad. 1993. Genotype, explant age and explant type as effecting callus and shoot regeneration in tomato. *Assiut J. Agri. Sci.*, **24**:3–14.

Ezura, H., S. Nishimiya and M. Kasumi. 1993. *Plant Cell Rep.*, **12**:676–680.

Fallon, K.M. and R. Phillips. 1989. Responses to water stress in adapted and undapted carrot cell suspension cultures. *J. Expt. Bot.*, **40(215)**:681–687.

FAO Statistical Database(2005)FAOSTATAgriculturedata,URL,http://apps. fao. org/collections?subset=agriculture.

Fari, M. and A. Andrasfalvy. 1994. Regeneration and cloning of pepper (*Capsicum* sp.): A review. *Hort. Sci.*, **26**:9–27.

Fari, M. and G. Csillery. 1983. *In vitro* shoot tip culture of some *Capsicum* spp. *Capsicum Newsletter*, **2**:70–73.

Fari, M. and G. Csillery. 1984. *In vitro* grafting of non-developed plants of *Capsicum cardenansis* hybrid produced by embryo culture. *Capsicum Newsletter*, **3**:51.

Fari, M. and M. Czako. 1981. Relationship between position and morphogenetic response of pepper hypocotyl explants cultured *in vitro*. *Sci. Hort.*, **15**:207–213.

Fari, M., A. Bank-Peredi and M.Toth-Gsanyi. 1991. *Acta Hortic.*, **289**:111.

Fari, M., A. Szasz, J. Mityko, I. Nagy, M. Csanyi and A. Andrasfalvy. 1992. Induced organogenesis via the seedling decapitation method (SDM) in three solanaceous vegetable species. *Capsicum Newsletter*, 243–248.

Fari, M., I. Nagy, M. Csanyi, J. Mityko and A. Andrasfalvy. 1995b. *Agrobacteirum*-mediated genetic transformation and plant regeneration via organogenesis and somatic embryogenesis from cotyledon leaves in egg plant (*Solanum melongena* L. cv. kecskemet; lila). *Plant Cell Rep.*, **15**:82–86.

Fari, M., M. Csanyi, J. Mityko, A. Peredi, A. Szasz and A. Csillag. 1995a. An alternative pathway of *in vitro* organogenesis in higher plants: plant regeneration via decapitated hypocotyls in three solanaceous vegetables genera. *Hort. Sci.*, **27**:9–16.

Fari, M., M. Laslo and J. Zatyko. 1982. Activity of endogenous cytokinins and the organisation of adventitious shoots in *in vitro* hypocotyl explants of pepper seedlings. *Capsicum Newsletters*, **1**:16–17.

Fari. M. 1983. Histological analysis of adventitious shoot bud formation in pepper (*Capsicum annuum* L.) hypocotyl explants cultured *in vitro*. *Capsicum Newsletter*, **2**:69–71.

Faria, R.T. and R.D. Illg. 1996. Inheritance of *in vitro* plant regeneration ability in the tomato. *Braz. J. Genet.*, **19**:113–116.

Faria, R.Td, D. Destro, J.C. Bespalhok Filho and R.D. Illg. 2002. Introgression of *in vitro* regeneration capability of *Lycopersicon pimpinellifolium* Mill. into recalcitrant tomato cultivars. Euphytica, **124**:59–63.

Filipone, E. and P.F. Lurquin. 1989. Stable transformation of eggplant (*Solanum melongena* L.) by co-cultivation of tissue with *Agrobacterium tumefasciens* carrying a binary plasmid vector. *Plant Cell Rep.*, **8**:370–373.

Fillatti, J.J., J. Kiser, R. Rose and L. Comai. 1987. Efficient transfer of a glyphosate tolerance gene into tomato using a binary *Agrobacterium tumefaciens* vector. *Bio Technol.*, **5**:726–730.

Fobert, P.K. and D.T. Webb. 1988. Effect of polyamines, polyamine precursors and polyamine inhibitors on somatic embryogenesis from eggplant (*Solanum melongena* L.) cotyledons. *Can J. Bot.*, **66**:1734–1742.

Fourvier, D., F. Lejeune and Y. Tourte 1995. Cytological events during the initiation of meristematic nodules in calli derived from eggplant nodules. *Biol. Cell.*, **85**:93–100.

Frankenberger, E.A., P.M. Hasegawa and E.C. Tigchelaar 1981a. Influence of environment and developmental state on the shoot forming capacity of tomato genotypes.

Z. Pflanzenphysiol., **102**:221–232.

Frankenberger, E.A., P.M. Hasegawa and E.C. Tigchelaar 1981b. Diallel analysis of shoot-forming capacity among selected tomato genotypes. *Z. Pflanzenphysiol.*, **102**:233–242.

Frary, A. and E.D. Earle 1996. An examination of factors affecting the efficiency of *Agrobacterium*-mediated transformation of tomato. *Plant Cell Rep.*, **16**:235–240.

Frijters, A., G. Simons, G. Varga, N. Quaedvlieg and M. de Both. 2000. The Mi-I gene confers resistance to the root-knot mematode *Meloidogyne incognita* in transgenic eggplant. VIII conf. Plant and animal genome, pp. 9–12, San Diego, C.A., USA.

Galmarini, C.R. 1997. Pepper breeding in Argentina. *Capsicum and Eggplant Newsletter*, **16**:28–34.

Geetha, N., P. Venkatachalam, P.S. Reddy and G. Rajaseger. 1998. *In vitro* plant regeneration from leaf callus cultures of tomato (*Lycopersicon esculentum* Mill.). Adv. *Plant Sci.*, **11**:253–257

Gill, R., K.A. Malik, M.H.M. Sanago and P.K. Saxena. 1995. Somatic embryogenesis and plant regeneration from seedling cultures of tomato (*Lycopersicon esculentum* Mill.). *J. Plant Physiol.*, **147**:273–276.

Gleba, Y.Y., N.M. Piven, I.K. Nomarnitski and N.K. Sytnik. 1984. Transmission genetics of the somatic hybridisation process in *Nicotiana*. 1. Hybrids and cybrids among the regenerants from cloned protoplast fusion products. *Theor. Appl. Genet.*, **69**:121–128.

Gleddie, S., W. Keller and G. Setterfield. 1983. Somatic embryogenesis and plant regeneration from leaf explants and cell suspensions of *Solanum melongena* (egg plant). *Can J. Bot.*, **61**:656–666.

Gleddie, S., W. Keller and G. Setterfield. 1986. Production and characterization of somatic hybrids between *Solanum melogena* L. and *S. sisymbrifolium* Lam. *Theor. Appl. Genet.*, **71**:613–621.

Gresshoff, P.M. and C.H. Doy. 1972. Development and differentiation of haploid *Lycopersicon esculentum* (tomato). *Planta*, **107**:161–170.

Grubben, G.J.M. 1977. Tropical vegetables and their genetic resources. *In*: M.D. Tindall and J.T. Williams (eds.), Vol 23 (pp. 34–37). IBPGR, Rome.

Gunay, A.L. and P.S. Rao. 1978. *In vitro* plant regeneration from hypocotyl and cotyledon explants of red pepper (*Capsicum*). *Plant Sci Lett.*, **11**:365–372.

Gunay, A.L. and P.S. Rao. 1980. *In vitro* propagation of hybrid tomato plants (*Lycopersicon esculentum* L.) using hypocotyl and cotyledon explants. *Ann. Bot.*, **45**:205–207.

Guri, A. and K.C. Sink. 1988a. Interspecific somatic hybrid plants between eggplant (*Solanum melongena*) and *Solanum torvum*. *Theor. Appl. Genet.*, **76**:490–496.

Guri, A. and K.C. Sink. 1988b. Organelle composition in somatic hybrids between an atrazine resistant biotype of *Solanum nigrum* and *Solanum melongena*. *Plant Sci.*, **58**:51–58.

Guri, A. and K.C. Sink. 1988c. *Agrobacterium* transformation of eggplant. *J. Plant Physiol.*, **133**:52–55.

Guri, A., L.J. Dunbar and K.C. Sink. 1991. Somatic hybridization between selected *Lycopersicon* and *Solanum* species. *Plant Cell Rep.*, **10**:76–80.

Haldrup, A., S.G. Petersen and F.T. Okkels. 1998. Positive selection: a plant selection principle based on xylose isomerase, an enzyme used in the food industry. *Plant Cell Rep.*, **18**:76–81.

Hamza, S. and Y. Chupeau. 1993. Re-evaluation of conditions for plant regeneration and *Agrobacterium*-mediated transformation from tomato (*Lycopersicon esculentum*). *J. Exp. Bot.*, **44**:1837–1845.

Harini, I. and G. Lakshmi Sita. 1993. Direct somatic embryogenesis and plant regeneration from immature embryos of chilli (*Capsicum annuum* L.). *Plant Sci.*, **89**:107–112.

Hendy, H., E. Pochard and A. Dalmarso 1985. Transmission de la resistance aux nematodes *Meloidogyne* chitwood (Tylenchida) Portee par 2 lignees de. *Capsicum annuum* L. Etude des descendance homozygoes issues d' androgenese. *Agronomie*, **5**:93–100.

Hobson, G. and J. Davies. 1971. The Tomato. *In*: A. Hulme (ed.) The Biochemistry of Fruits and their Products (pp. 337–482). Academic Press, New York, London.

Hwang, J.K., K.Y. Paek and C.H. Cho. 1998. *Acta Hortic.*, **461**:301–307.

Iannacone, R., M.C. Fiore, A. Macchi, P.D. Grieco, S. Arpaia, D. Perrone, G. Mennella, F. Sunseri, F. Cellini and G.L. Rotino. 1995. Genetic engineering of eggplant (*Solanum melongena* L.) *Acta Horti.*, **392**:227–233.

Ichimura, K. and M. Oda. 1998. Stimulation of phenotypically normal shoot regeneration of tomato (*Lycopersicon esculentum* Mill.) by commercial filter paper extract. *J. Jap. Soc. Hort. Sci.*, **67**:378–380.

Imanishi, S., S. Masataka and I. Hiura. 1976. Organogenesis from long term callus culture from hypocotyle in *Lycopersicon esculentum* L. *Agric. Sci.*, **7**:439–450.

Ionescu, A. 1998. Anale institutul de Cercetari Pentru Legumicultura si Floricultura Vidra, **15**:17–27.

Isouard, G., C. Raquin and Y. Demarly. 1979. Obtention de plantes haploïdes et diploïdes par culture *in vitro* d'anthères d'aubergine (*Solanum melongena* L.). *C.R. Acad. Sci.* Paris, **288**:987–989.

Izadpanah, M. and M. Khosh-Khui. 1992. Comparisons of *in vitro* propagation of tomato cultivars. *Iran Agric. Res.*, **8**:37–47.

Izvorska, N.D., G.N. Vassilev and D.T. Lilov. 1982. Influence of the cytokinins p-allyl and phenylthioureidosalicylic acids on the growth and morphogenesis of callus tissues from vine and tomato. *Comptes Rendus de L'Academie Bulgare des Sci.*, **35**:387–390.

Jain, S.M., E.A. Shahin and S. Sun. 1988. Interspecific protoplast fusion for the transfer of atrazine resistance from *Solanum nigrum* to tomato (*Lycopersicon esculentum* L.). *Plant Cell Tiss. Org. Cult.*, **12**:189–192.

Janick, J. 1993. Genetically engineered high solid tomat:o. *In* New Crops News, Spring. http://www.hort.purdue.edu/newcrop/ NewCropsNews/93-3-/tomato.html. Date of access 20 June 2003.

Jaramillo, J. and W.L. Summers. 1990. *J. Amer. Soc. Hort. Sci.*, **115**:1047–1050.

Jaramillo, J. and W.L. Summers. 1991. *HortScience*, **26**:915–916.

Jarl, C.I., E.M. Rietveld and J.M. De Haas. 1999. Transfer of fungal tolerance through interspecific somatic hybridisation between *Solanum melongena* and *S. torvum*. *Plant Cell Rep.*, **18**:791–796.

Jawahar, M., S.V. Mohamed and N. Jayabalan. 1997. *In vitro* callus culture and plant regeneration from different explants of *Lycopersicon esculentum* Mill. *J. Phytol. Res.*, **10**:75–78.

Jelenkovic, G., S. Billings, Q. Chen, Lashomb, G. Hamilton and G. Ghidiu. 1998. Transformation of eggplant with synthetic CryIIIA gene produces a high level of resistance to the Colorado potato beetle. *J. Am. Soc. Hort. Sci.*, **123**:19–25.

Jia, J. and I. Potrykus. 1981. Mesophyll protoplasts from *Solanum melongena* var. *depressum* Bailey regenerate to fertile plants. *Plant Cell Rep.*, **1**:71–72.

Jiang, Z.R. and C.L. Li. 1984. Observations and experiments on late generation of sweet and hot pepper derived by anther culture. *Acta Hort. Sinica*, **11**:191–194.

Joersbo, M., I. Donaldson, J. Kreiberg, S.G. Petersen, J. Brunstedt and F.T. Okkels. 1998. Analysis of mannose selection used for transformation of sugar beet. *Molecular Breeding*, **4**:111–117.

Johnson, T.S., R. Sarada and G.A. Ravishankar. 1998. Capsicin formation in p-fluorophenylalanine resistant and normal cell cultures of *Capsicum frutescens* and activity of phenylalanine ammonia lyase. *J. Biosci.*, **23**:209–212.

Jones, M. 1988. Fusing plants protoplasts. *TibTech.*, **6**:153–158.

Kalloo, G. 1991. Introduction. *In*: G. Kalloo (eds.) Monographs on Theoretical and Applied Genetics 14, Genetic Improvement of Tomato (pp. 1–9). Springer-Verlag, Berlin, Heidelberg, New York.

Kamat, M.G. and N.A. Rao. 1978. Vegetative multiplication of eggplants (*Solanum melongena*) using tissue culture techniques. *Plant Sci. Lett.*, **13**:57–65.

Kaparakis, G. and P.G. Alderson. 2002. Influence of high concentrations of cytokinins on the production of somatic embryos by germinating seeds of tomato, aubergine and pepper. *J. Hort. Sci. Biotechnol.*, **77**:186–190.

Kartha, K.K., O.L. Gamborg, J.P. Shyluk and F. Constabel. 1976. Morphogenetic investigations on *in vitro* leaf culture of tomato (*Lycopersicon esculentum* Mill. cv. Starfire) and high frequency plant regeneration. *Z. Pflanzenphysiol.*, **77**:292–301.

Kaul, M. 1991. Reproductive biology of tomato. *In*: G. Kalloo (ed.) Monographs on Theoretical and Applied Genetics 14, Genetic Improvement of Tomato (pp. 39–43). Springer-Verlag, Berlin, Heidelberg, New York.

Khan, R. 1979. *Solanum melongena* and its ancestral forms. *In*: J.C. Hawkes, J.G. Lester and A.D. Skelding (eds.) The biology and taxonomy of the *Solanaceae*, (pp. 629–638). Linnean Soc., Academic Press, London.

Kim, Y.H., J.K. Kim, Y.S. Hwang and K.H. Pack. 1995. *In*: Proc. Induced Mutations and Molecular Techniques for crop Imporvement, Vienna, Austria, June 19–23, 1995, pp. 147–155.

Kintzois, S., J.B. Drossopoulos and Ch. Lymperopoulos. 2001. Effect of vitamins and inorganic micronutrients on callus growth and somatic embryogenesis from leaves of chilli pepper. *Plant Cell Tiss. Org. Cult.*, **67**:55–62.

Kombrink, E. and I.E. Sonsich. 1995. Defence responses of plants to pathogens. *Adv. Bot. Res.*, **25**:1–34.

Koornneef, M., J. Bade, C. Hanhart, K. Horsman, J. Schel, W. Soppe, R. Verkerk and P. Zabel. 1993. Characterization and mapping of a gene controlling shoot regeneration in tomato. *Plant J.*, **3**:131–141.

Kumar, P.A., A. Mandaokar, K. Sreenivasu, S.K. Chakrabarti, S. Bisaria, S.R. Sharma, S. Kaur and R.P. Sharma. 1998a. Insect resistant transgenic brinjal plants. *Mol. Breeding*, **4(1)**:33–37.

Kumar, P.A., A.D. Mandaokar and R.P. Sharma. 1998b. Genetic engineering for the improvement of eggplant (*Solanum melongena* L.). *AgBiotech.*, **10**:329–332.

Kurtz, S.M. and R.D. Lineberger. 1983. Genotypic differences in morphogenic capacity of cultured leaf explants of tomato. *J. Am. Soc. Hort. Sci.*, **108**:710–714.

Lech, M., K. Miczynski and A. Pindel 1996. Comparison of regeneration potentials in tissue cultures of primitive and cultivated tomato species (*Lycopersicon* sp.). *Acta Soc. Bot. Poloniae*, **65**:53–56.

Lee, H.Y., J.O. Jeon, J.G. No and H.G. Park. 1999. *Horti. Sci.*, **40**:537–540.

Lee, Y.H., H.S. Kim, J.Y. Kim, M. Jung, Y.S. Park, J.S. Lee, S.H. Choi, N.H. Her, J.H. Lee, N.I. Hyung, C.H. Lee, S.G. Yang and C.H. Harn. 2004. A new selection method for pepper transformation:callus-mediated shoot formation. *Plant Cell Rep.*, **23**:50–58.

Leone, M., E. Filippone and P.F. Lurquin. 1993. Transformation in *Solanum melongena* L. (eggplant). *In*: Bajaj YPS. (ed.) Biotechnology in Agriculture and Forestry: Plant Protoplasts and Genetic Engineering (pp. 320–328). Springer-Verlag, Berlin, Germany.

Li, C.J., B.J. Ye and D.W. Zhu. 1995. *Acta Hortic.*, **402**:442–444.

Li, D., K. Zhao, B. Xie, B. Zhang and K. Luo. 2003. Establishment of highly efficient transformation system for pepper (*Capsicum annuum* L.). *Plant Cell Rep.*, **21**:785–788.

Lim, H.T., G.Y. Lee, Y.S. You, E.J. Park, Y.N. Song, D.C. Yang and K.H. Choi. 1999. *Acta Hortic.*, **483**:387–396.

Liu, K.B., Y.M. Li and K.C. Sink. 1995. Asymmetric somatic hybrid plants between an interspecific *Lycopersicon* hybrid and *Solanum melongena*. *Plant Cell Rep.*, **14**:652–656.

Liu, W., W.A. Parrott, D.F. Hildebrand, G.B. Collins and E.G. Williams. 1990. *Agrobacterium*-induced gall formation in bell pepper (*Capsicum annuum* L.) and formation of shoot-like structures expressing introduced genes. *Plant Cell Rep.*, **9**:360–364.

Lukyanenko, A.N. 1991. Disease resistance in tomato. *In*: G. Kalloo (ed.) Monographs on Theoretical and Applied Genetics 14, Genetic Improvement of Tomato (pp. 99–119). Springer-Verlag, Berlin, Heidelberg, New York.

Ma, Y.H., K. Kato Kato and M. Msuda 1999. *J. Jap Soc. Hort. Sci.*, **68**:768–773.

Magioli, C., A.P.M. D.E. Rocha de Oliveira and E. Mansu. 1998. Efficient shoot organogenesis of eggplant (*Solanum melongena* L.) induced by thidiazuron. *Plant Cell Rep.*, **17**:661–663 .

Manoharan, M., C.S.S. Vidya and G.L. Sita. 1998. *Agrobacterium*-mediated genetic transformation in hot chilli (*Capsicum annuum* L. var Pusa Jwala). *Plant Sci.*, **131**:77–83.

Matsuoka, H. and K. Hinata. 1979. NAA-induced organosgenesis and embryogenesis in hypocotyl callus of *Solanum melongena* L. *J. Exp. Bot.*, **30**:363–370.

Mensuali-Sodi, A., M. Panizza and F. Tognoni. 1995. Endogenous ethylene requirement for adventitious root induction and growth in tomato cotyledons and lavandin microcuttings *in vitro*. *Plant Growth Reg.*, **17**:205–212

Miyoshi, K. 1996. Callus induction and plantlet formation through culture of isolated microspores of eggplant (*Solanum melongena* L.). *Plant Cell Rep.*, **15**:391–395.

Momiyama, T., J.C. Afele, T. Saito, T. Kayano, Y. Tabei, F. Takaiwa, K. Takayanagi and S. Nishimura. 1995. *Biochem. Biophys. Res. Comm.*, **213**:376–382.

Morrison, R.A. 1987. Dissertation Abst. *International B. Sci. Eng.*, **48**:187.

Muhlbach, H.P. 1980. Different regeneration potentials of mesophyll protoplasts from cultivated and a wild species of tomato. *Planta*, **148**:89–96

Munyon, I.P., J.F. Hubstenberger and G.C. Phillips. 1989. Origin of plantlets and callus obtained from chilli pepper anther cultures. *In vitro Cell and Dev Biol.*, **25**:293–296.

Newman, P.O., S. Krishnaraj and P.K. Saxena. 1996. Regeneration of tomato (*Lycopersicon esculentum* Mill.): somatic embryogenesis and shoot organogenesis from hypocotyl explants induced with 6-benzyladenine. *Int. J. Plant Sci.*, **157**:554–560.

Nguyen, H.M., N.N. Balashova, P.K. Kintya, V.A. Smirnov and T.T. Kieu. 1992. Effects of steroid glycosides on tomato tissue culture. *Sel'Skokhozyaistvennaya Biologiya*, pp. 57–63.

Oankh, Kh K., N. Zagorska and H. Georgiev. 1986. *Genetika I Selektisya*, **19**:425–427.

Ochoa-Alejo, N. and R. Ramirez-Malagon. 2001. *In vitro* chili pepper Biotechnology. *In Vitro Cell Dev. Biol.* (Plant), **37(6)**:701–729.

Ohki, S., C. Bigot and J. Mousseau. 1978. Analysis of shoot forming capacity *in vitro* in two lines of tomato (*Lycopersicon esculentum* Mill.) and their hybrids. *Plant Cell Physiol.*, **19**:27–42.

Pelletier, G., F. Vedel and G. Belliard 1985. Cybrids in genetics and breeding. *Hereditas*.

Peres, L.E.P., P.G. Morgante, C. Vecchi, J.E. Kraus and M.A.V. Sluys 2001. Shoot regeneration capacity from roots and transgenic hairy roots of tomato cultivars and wild related species. *Plant Cell Tiss. Org. Cult.*, **65**:37–44.

Peter, K.V., T.R. Gopalakrishnan, S. Rajan and P.G. Sadhan Kumar. 1993. *In*: G.L. Hartman and A.C. Haywards (eds.) Bacterial wilt, ACIAR Proc. 45, Oct. 28–31 (pp. 183–190).

Phillips, G.C. and J.F. Hubstenberger 1985. Organogenesis in peper tissue culture. *Plant Cell Tiss Org Cult.*, **4**:262–269.

Pijnacker, L.P., M.A. Ferwerda, K.J. Puite and J.G. Schaart. 1989. Chromosome elimination and mutation in tetraploid somatic hybrids of *Solanum tuberosum* and *Solanum phure Ja. Plant Cell Rep.*, **8**:82–85.

Plastira, VA. and A.K. Perdikaris. 1997. Effect of genotype and explant type in regeneration frequency of tomato *in vitro*. *Acta Horti.*, pp. 231–234.

Polisetty, R., S. Khaterpal and P. Patil. 1994. *Seed. Res.*, **22**:26–29.

Prakash, A.H., KS. Rao and M. Udaya Kumar. 1997. Plant regeneration from protoplasts of *Capsicum annuum* L. cv california wonder. *J. Biosci.*, **22**:339–344.

Purseglove, J.W. 1977. Tropica crops Dicotyledons I and II. Longman, London. p. 524.

Ramulu, K. and M. Denereux. 1979. *Incompatibility Newslett.*, **11**:68–71.

Rao, A. and S. Agarwal. 2000. Role of antioxidant lycopene in cancer and heart disease. *J. Am. College Nutr.*, **19**:563–569.

Rhodes, D. 2002.TomatoesNotes(PurdueUniversity)//www. hort. purdue. edu/rh/ rhodccv/hort410/tomato/to00001. htm.

Romer, S., P.D. Fraser, J.W. Kiano, C.A. Shipton, N. Misawa, W. Schuch and P.M. Bramley. 2000. Elevation of the provitamin A content of transgenic tomato plants. *Nature Biotechnol.*, **18**:666–669.

Rotino, G.L. and S. Gleddie. 1990. Transformation of eggplant (*Solanum melongena* L.) using a binary *Agrobacterium tumefasciens* vector. *Plant Cell Rep.*, **9**:26–29.

Rotino, G.L., E. Perri, A. D'Alessandro and G. Mennella. 1998. Characterisation of fertile somatic hybrids between eggplant (*S. melongena* L.) and *S. integrifolium*. *In*: Proc. Xth Eucarpia meeting on Genetics and Breeding of Capsicum and Eggplant, (pp. 213–217). Avignon, France.

Rotino, G.L., E. Perri, M. Zottini, H. Sommer and A. Spena. 1997. Genetic engineering of parthenocarpic plants. *Nat. Biotech.*, **15**:1398–1401.

Rotino, G.L., E. Perri, N. Acciarri, F. Sunseri and S. Arpaia. 1997. Development of eggplant varietal resistance to insects and diseases via plant breeding. *Adv. Hort. Sci.*, **11**:193–201.

Rotino, G.L., Pedrazzini, Dorr I. and E. Perri. 1995. Fusione di protoplasti tra melazana (*Solanum melongena* L.) e *Solanum selvatici*. *In*: Proc. XXXIX, *Convegno Annuale S.I.G.A.*, (pp. 157), Vasto Marina, Italia.

Saito, T. 1999. *Bull. Natl. Res. Inst. Veg. Ornamental Pl. and Tea*, **14**:105–164.

Saito, T. and S. Nishimura. 1994. *Plant Sci. Limer.*, **102**:205–211.

Samoylov, V.M. and K.C. Sink. 1996. The role of irradiation dose and DNA content of somatic hybrid calli in producing asymmetric plants between an interspecific tomato hybrid and eggplant. *Theor. Appl. Genet.*, **92**:850–857.

Samoylov, V.M., S. Izhar and K.C. Sink 1996. Donor chromosome elimination and organelle composition of asymmetric somatic hybrid plants between an interspecific tomato hybrid and eggplant. *Theor. Appl. Genet.*, **93**:268–274.

Santana, N. 1985. *Cultivos Tropicales*, **7**:121–128.

Saxena, P.K., R. Gill, A. Rashid and S.C. Maheshwari. 1981. Isolation and culture of protoplasts of *Capsicum annuum* L. and their regeneration into plants flowering *in vitro*. *Protoplasma*, **108**:357–360.

Schuler, T.H., G.M. Poppy, B.R. Kerry and I. Denholm 1998. Insect resistance transgenic plants. *Trends Biotechn.*, **16**:168–175.

Schutze, R. and G. Wieczorrek. 1987. Investigations into tomato tissue cultures. I. Shoot regeneration in primary explants of tomato. *Arch. Zuchtungsforschung*, **17**:3–15.

Selassie, K.G., E. Pochard, G. Marchoux and J.C. Thouvene. 1986. VIth meeting on Genetics and Breeding of *Capsicum* and Eggplant (pp. 189–192), Zargoza (Spain) October 21–24, 1986.

Shahin, E.A. 1985. Totipotency of tomato protoplasts. *Theor. Appl. Genet.*, **69**:235–240.

Sharma, P. and M.V. Rajam. 1995a. Genotype, explant and position effects on organogenesis and somatic embryogenesis in eggplant (*Solanum melongena* L.). *J. Exp. Bot.*, **46**:135–141.

Sharma, P. and M.V. Rajam. 1995b. Spatial and temporal changes in endogenous polyamine levels associated with somatic embryogenesis from different hypocotyl segments of eggplant (*Solanum melongena* L.). *J. Plant Physiol.*, **146**:658–664.

Shcehreva, L., M. Kruleva and N. Zagorska. 1990. *Genetika I Selektisya*, **23**:185–193.

Shifriss, C., M. Pilowsky and J.M. Zacks. 1992. *Capsicum Newslett, Special Issue*, pp. 172–177.

Sihachakr, D. and G. Ducreux 1987. Cultural behavior of protoplasts from different organs of eggplant (*Solanum melongena* L.) and plant regeneration. *Plant Cell Tiss. Org. Cult.*, **11**:179–188.

Sihachakr, D., G. Ducreux, F. Vedel, M. Allot, L.H. San and A. Server 1989b. Somatic hybridization of eggplant (*Solanum melongena* L.) with *Solanum nigrum* L. by protoplast electrofusion. Int. Conf. The Impact of Biotechnology on Agriculture, 10–12 July, 1989, Arniens, France.

Sihachakr, D., M.C. Daunay, I. Serraf, M.H. Chaput, I. Mussio, R. Haicour, L. Rossignol and G. Ducreux. 1994. Somatic hybridization of eggplant (*Solanum melongena* L.) with its close and wild relatives. *In*: Y.P.S. Bajaj (ed.) Biotechnology in Agriculture and Forestry, Somatic Hybridization in Crop Improvement, Vol I (pp. 255–278). Springer-Verlag, Berlin, Heidelberg.

Sihachakr, D., M.H. Chaput, I. Serraf and G. Ducreux. 1993. Regeneration of plants from protoplasts of eggplant (*Solanum melongena* L.). *In*: Y.P.S. Bajaj (ed.) Biotechnology in Agriculture and Forestry, Plant Protoplasts and Genetic Engineering, Vol IV (pp. 108–122). Springer-Verlag, Berlin, Heidelberg.

Sihachakr, D., R. Haicour, E. Barrientos, G. Ducreux and L. Rossignol. 1989a. Somatic hybrid plants produced by electrofusion between *Solanum melongena* L. and *Solanum torvum* Sw. *Theor. Appl. Genet.*, **77**:1–6.

Sihachakr, D., R. Haicour, I. Serraf, E. Barrientos, C. Herbreteau, G. Ducreux, L. Rossignol and V. Souvannavong. 1988. Electrofusion for the production of somatic hybrid plants of *Solanum melongena* L. and *Solanum khasianum* C.B. Clarke. *Plant Sci.*, **57**:215–223.

Sink, K.C., L.W. Handley, R.P. Niedz and P.P. Moore. 1986. Protoplast culture and use of regeneration attributes to select tomato plants. *Genet. Manip.* pp. 405–413.

Sireagar, E.B.M. and Sudersono. 1997. Shoot regeneration from hypocotyl segments of hot pepper mediated by non-disarmed isolates of *Agrobacterium*. *Capsicum and Eggplant News Letters*, **16**:102–105.

Sripichitt, P., E. Nawata and S. Shigenaga 1987. *In vitro* shoot forming capacity of cotyledon explants of red pepper (*Capsicum annuum* L. cv Yatsufusa). *Jpn J. Breed.*, **37(2)**:133–142.

Steinitz, B., D. Wolf, J. Matzevitch-Josef and A. Zelcer 1999. Regeneration *in vitro* and genetic transformation of pepper (*Capsicum* spp.). *The current state of art. Capsicum and Eggplant Newsletter*, **18**:9–15.

Stommel, J.R. and S.L. Sinden. 1991. Genotypic differences in shoot forming capacity of cultured leaf explants of *Lycopersicon hirsutum*. *HortScience*, **26**:1317–1320.

Stone, B. 1994. The Flavr Savr arrives. http://www. accessexcellence. org/AB/BA/Flavr_Savr_Arrives. html, Date of access 25 June 2003.

Subhash, K. and B. Prolaram 1987. Induction of multiple shoots by gamma rays from cotyledon cultures of *Capsicum*. *Curr Sci.*, **56**:1289–1291.

Subhash, K. and T. Christopher. 1988. Direct plantlet formation in cotyledon cultures of *Capsicum frutescens*. *Curr. Sci.*, **57**:99–100.

Summers, W.L., Jaramillo and T. Baily. 1992. *HortScience*, **27**:838–840.

Sunseri, F., M.C. Fiore, F. Mastrovitro, E. Tramontano and G.L. Rotino. 1993. *In vitro* selection and genetic analysis for kanamycin resistance in transgenic eggplant (*Solanum melongena* L.). *J. Genet. Breed.*, **47**:299–306.

Takashina, T., T. Suzuki, H. Egashira and S. Imanishi. 1998. New molecular markers linked with the high shoot regeneration capacity of the wild tomato species *Lycopersicon chilense*. *Breed. Sci.*, **48**:109–113.

Tal, M., K. Dehan and H. Heikin. 1977. Morphogenetic potential of cultured leaf sections of cultivated and wild species of tomato. *Ann. Bot.*, **41**:937–941.

Toki, S., T. Kameya and T. Abe. 1990. Production of a triple mutant, chlorophyll-deficient, streptomycin-and kanamycin-resistant *Nicotiana tabacum*, and its use in intergeneric somatic hybrid formation with *Solanum melongena*. *Theor. Appl. Genet.*, **80**:588–592.

Valera-Montero, L.L. and N. Ochoa-Alejo 1992. A novel approach for chili pepper (*Capsicum annuum* L.) Plant regeneration: Shoot induction in rooted hypocotyls. *Plant Sci.*, **84**:215–219.

Vaquero, C., M. Sack, J. Chandler, J. Drossard, F. Schuster, M. Monecke, S. Schillberg and R. Fischer. 1999. Transient expression of a tumor-specific single-chain fragment and chimeric antibody in tobacco leaves. *Proc. Natl. Acad. Sci.*, **96**:11128–11133.

Vaulx, R.D. de, E. Pochard and D. Chambonnet. 1982. *Capsicum Newslett.*, **1**:52–53.

Venczel, G. and J. Mityko. 1995. *Zoldsegtemesztesi Kutato Intezet Bulletinje*, **27**:145–156.

Venkatachalam, P., N. Geetha, P. Priya, G. Rajaseger and N. Jayabalan. 2000. High frequency plantlet regeneration from hypocotyls explants of tomato (*Lycopersicon esculentum* Mill.) via organogenesis. *Plant Cell Biotech. Mol. Biol.*, **1**:95–100.

Venkataiah, P. and K. Subhash. 2001. Genotype, explant and medium effects on adventitious shoot bud formation and plant regeneration of red pepper (*Capsicum annuum* L.). *Journal Genetics and Breeding*, **55**:143–149.

Venkataiah, P., T. Christopher and K. Subhash 2001. Plant regeneration and *Agrobacterium* mediated genetic transformation in four *Capsicum* species. *Capsicum and Eggplant Newsletter*, **20**:68–71.

Venkataiah, P., T. Christopher and K. Subhash 2003. Thidiazuron-induced adventitious shoot bud formation and plant regeneration *Capsicum annuum* L. *Journal of Plant Biotechnology*, **5(4)**:245–250.

Villiers, R.P.D., R.J.V. Vuuren, D.I. Ferreira and J.V. Staden. 1993. Regeneration of adventitious buds from leaf discs of *Lycopersicon esculentum* cv. Rodade: optimization of culture medium and growth conditions. *J. South African Soc. Hort. Sci.*, **3**:24–27.

Vnuchkova, V.A. 1977a. Development of a method for obtaining regenerate tomato plants under tissue culture conditions. *Fiziol. Rast.*, **24**:1094–1100.

Vnuchkova, V.A. 1977b. Elaboration of methods for obtaining tomato plants by tissue culture. *Fiziol. Rast.*, **24**:1095–1100.

Wang, Y., M. Yang, N. Pan and Z. Chen. 1991. Plant regeneration and transformation of Sweet pepper (*Capsicum frutescens*). *Acta. Bot. Sin.*, **33**:780–786.

Wijbrandi, J., J.G.M. Vos and M. Koornneef. 1988. Transfer of regeneration capacity from *Lycopersicon peruvianum* to *L. esculentum* by protoplast fusion. *Plant Cell Tiss. Org. Cult.*, **12**:193–196.

Wijbrandi, J., W.V. Capelle, C.J. Hanhart, E.P.V.L. Martinet-Schuringa and M. Koornneef.1990. Selection and characterization of somatic hybrids between *Lycopersicon esculentum* and *Lycopersicon peruvianum*. *Plant Sci.*, **70**:197–208.

Yadav, J.S. and M.V. Rajam 1997. Spatial distribution of free and conjugated polyamines in leaves of *Solanum melongena* L. associated with differential morphogenetic capacity: efficient somatic embryogenesis with putrescine. *J. Exp. Bot.*, **48**:1537–1545.

Yadav, J.S. and M.V. Rajam 1998. Temporal regulation of somatic embryogenesis by adjusting cellular polyamine content in eggplant. *Plant Physiol.*, **116**:617–625.

Yamada, T., H. Nakagawa and Y. Sinoto. 1967. *Bot Mag. Tokyo*, **80**:68–74.

Yamamoto, Y. and O. Matsumoto. 1998. *J. Jap. Soc. Hort. Sci.*, **63**:67–72.

Zagorska, N.A., A. Shtereva, B.D. Dimitrov and M.M. Kruleva. 1998. Induced androgenesis in tomato (*Lycopersicon esculentum* Mill.). I. Influence of genotype on androgenetic ability. *Plant Cell Rep.*, **17**:968–973.

Zagorska, N.A., M.D. Abadjieva and H.A. Georgiev. 1982. Inducing regeneration in anther cultures of tomatoes (*Lycopersicon esculentum* Mill.). *Comptes Rendus de L'Academie Bulgare des Sci.*, **35**:97–100.

Zamit, D., R.A. Jones and H.K. Oanh. 1980. *Plant Sci. Lett.*, **17**:353–361.

Zamorzaeva, I.A. and V.A. Smirnov. 1987. Izvestiya Akademii Nauk Moldavskai SSR, *Biologichenskikh I Khimicheskikh Nauk*, **1**:31–34.

Zapata, F.J. 1980. Different regeneration potentials of mesophyll protoplasts from cultivated and a wild species of tomato. *Planta*, **148**:89–96.

Zeitlin, L., S.S. Olmsted, T.R. Moench, M.S. Co, B.J. Martinell, V.M. Paradkar, D.R. Russell, C. Queen, R.A. Cone and K.J. Whaley. 1998. A humanized monoclonal antibody produced in transgenic plants for immuno protection of the vagina against genital herpes. *Nat. Biotechnol.*, **16**:1361–1364.

Zelcer, A., O. Soferman and S. Izhar. 1983. Shoot regeneration in root cultures of *Solanaceae. Plant Cell Rep.*, **2**:252–254

Zhang, H. and E. Blumwald. 2001. Transgenic salt-tolerant tomato plants accumulate salt in foliage but not in fruit. *Nat. Biotechnol.*, **19**:765–768

Zhang, Z., Z. Zhou, Y. Liu, Q. Jiang, M. You, G. Liu, J. Mi, Z.J. Zhang, Z.X. Zhou, Y.J. Liu, Q.Y. Jiang, M. You, G.M. Liu and J.J. Mi. 1994. CMV Cp gene transformation into pepper and expression in the offspring of the transgenic. *Acta Agriculturae Boreari Sinica*, **9**:67–71.

Zhu, K., Y.F. Zhang Wen-Jun and Z.L. Chen Ou-Yang. 1996. Transgenic sweet pepper plants from *Agrobacterium*-mediated transfromation. *Plant Cell Rep*, **16**:71–75.

Zhu, Z.J., Z.X. Zhou, Y.J. Liu, Q.Y. Jiang, M. You, G.M. Liu and J.J. Mi. 1994. *Acta Boreli Sinica*, **9**:67–71.

In vitro Multiple Shoots Induction from Cotyledon Explants of *Solanum surattense* A Medicinal Herb

T. Ugandhar, M. Praveen, M. Upender, M. Rambabu, K. Subhash and *N. Rama Swamy**

ABSTRACT

Solanum surattense is an important multipurpose medicinal plant commonly used in most of the ayurvedic preparations. High frequency of adventitious direct shoot bud induction and plant regeneration was observed from cotyledon explants on MS medium supplemented with BAP and Kn in various concentrations individually and in combination with IAA (0.5 mg/l). Direct adventitious shoot regeneration on MS medium containing various concentrations of BAP (1.0–8.0 mg/l) was observed with varied results. Hightest responding cultures with maximum frequency of multiple shoot induction was observed at 3.0 mg/l BAP (29.3 shoots/explant) followed by 4.0 mg/l BAP (25.3 shoots/explant). Whereas at 3.0 mg/l Kn shoot induction was less compared to BAP (25.7 ± 0.35 explant) followed by 4.0 mg/l Kn in which profuse rhizogenesis was also observed. Micro shoots (3–4 cm) developed from cotyledon regenerants were excised and cultured on ½ strength MS medium supplemented with 0.5 mg/l IAA and rooted plantlets were transplanted to plastic pots containing vermiculite and perlite (3:1) and grown in culture room. High rate of survival percentage (75%) was achieved during plant establishment. This protocol can be used for large scale multiplication and genetic transformation experiments to transfer novel traits for improvement of the species.

Plant tissue culture technology has a great impact in the multiplication and conservation of medicinal and threatened plants and also production of plants with agronomically important traits for the crop improvement. This approach not only reduces the duration for micropropagation and

---

* Plant Biotechnology Research Group, Department of Botany, Kakatiya University, A.P. 506009, India. E-mail: swamynr_dr@yahoo.com

selection of desirable traits but also facilitates raising of a large number of plants within a limited time and space and also it is an essential prerequisite for genetic engineering and the production of transgenic plants.

The plantlets regenerated from callus cultures show genetic variability and deviate from the normal diploids, which lead to somaclonal variation. Therefore, direct shoot morphogenesis from the primary tissue (explant) is more desirable than via an intermediate callus phase (Larkin and Scrowcroft, 1981). In direct regeneration, shoots/shoot buds, arise directly from a plant organ or a piece of tissue without an intervening callus phase. There have been numerous reports in recent years on regeneration of plants from various types of excised explants, *viz.*, cotyledon, hypocotyl, leaf, root, embryos etc. of many species (Lee and Phillips, 1988; Subhash and Christopher, 1988; Swartz *et al.*, 1990; Desai and Mehta, 1990; Kunjumon *et al.*, 1996; Rajani and Urs, 1998; Jas Rai *et al.*, 1999; Leflay *et al.*, 1999; Jeong *et al.*, 2001; Neeta *et al.*, 2001).

In the present investigation the role of different plant growth regulators on direct regeneration of cotyledon explants was studied to find out the regeneration potentiality and also the importance of cytokinins and auxin-cytokinin combinations in *S. surattense.*

Solanum surattense is medicinally important (Gamble, 1986; Sivarajan and Balachandran, 1999) and it is extensively used in the treatment to cough, bronchitis, cholera, dysuria, diabetes and asthma. It is also useful in cases of influenza, enteric fever and allied conditions. The drug is diuretic, expectorant, febrifuge, cardiotonic, laxative, stimulant and is also used against difficult urination, bladder stones, rheumatism, sore-throat, enlargement of liver and spleen, vomiting and skin diseases (Sivarajan and Balachandran, 1999). Fumigation with the seeds is a reputed cure for toothache (Nadkarni, 1954). Roots, fruits and occasionally the whole plant are used in medicine. The important formulations using the drug are *"kantakari ghrtham, Putikaranjasavam, Suranadi leham* etc." (Sivarajan and Balachandran, 1999). The efficacy of *Kantakari* in the treatment of bronchial asthma (Shwas and Kas) and non-specific cough has been tried and also studied the pharmacognosy of this species. *S. surattense* is also used as one of the ingredients in the preparation of ayurvedic massage oil *"Kama virya"*. The species is used in ayurvedic medicine as *"Dasamula"* for chest pains, cough, asthma, fever etc. and fruit juice in earache (Nayar *et al.*, 1989). Although the species *S. surattense* has importance in ayruvedic medicine, a little work has been done *in vitro* (Prasad and Chaturvedi, 1978; Prasad *et al.*, 1998; Rama Swamy *et al.*, 2005). In view of its medicinal value and used as vegetable, an attempt has been made to establish the efficient protocol for direct multiple shoots induction from cotyledon explants in *S. surattense.*

Materials and methods

Seeds of *S. surattense* were soaked for 24 hrs in sterilized distilled water and surface sterilized with 0.1% $HgCl_2$ for 3 to 5 minutes. Then these were washed 3 times with sterile distilled water for 5 minutes and germinated aseptically on MS basal medium (Murashige and Skoog, 1962). Cotyledon explants (0.6–0.8 cm^2) from 2 week old seedlings were excised and cultured on MS medium containing 3% (w/v) sucrose with various concentrations of cytokinins BAP, Kn (1.0–8.0 mg/l) alone and also in combination with auxins IAA (0.5 mg/l) (Tables 1–3). The pH of the media was adjusted to 5.8 with 0.1N HC1 or 0.1N NaOH, solidified with 0.8% Difco-Bacto agar and autoclaved at 121°C under 15 psi for 15–20 minutes. Single explant was inoculated in each culture tube and incubated at 25 ± 2°C under white fluorescent light of 40–60 µmol $m^{-2} s^{-1}$ intensity for 16 hr light/8 hr dark period.

In vitro *rooting and acclimatization*

The regenerated micro shoots were excised and transferred on to rooting medium containing ½ strength MS basal medium fortified with 0.5 mg/l and 1.0 mg/l IAA. After 4 weeks, the number of rooted shoots, and the number of roots per shoot were recorded. Well rooted shoots were rinsed with sterile water to remove agar and transferred to plastic pots containing sterile vermiculite and perlite (1:1) and kept in culture room. After 2 weeks, the plants were transferred to earthenware pots consisting of garden soil and grown in green house.

Data analysis

20 replicates were maintained for each treatment. Each treatment was repeated at least twice with similar results. Data were recorded after 6 weeks of culture.

Results

Effect of various concentrations of cytokinins such as BAP and Kn alone and in combination with various concentrations of auxin IAA was studied on direct multiple shoot bud induction in cotyledon explants of *S. surattense* (Tables 1–3). Direct adventitious shoot regeneration on MS medium containing various concentrations of BAP (1.0–8.0 mg/l) was observed with varied results. Higher responding cultures with maximum frequency of multiple shoot bud induction was observed at 3.0 mg/l BAP (29.0 shoots/explant) followed by 4.0 mg/l BAP (25.0 shoots/explant). At high concentration of BAP the percentage of responding cultures and shoot bud proliferation were reduced (Table 1).

Table 1: Direct shoots formation from cotyledon explants of *S. surattense* cultured on MS medium supplemented with different concentrations of BAP/Kn

	Concentration of growth regulators (mg/l)	*% of cultures responding*	*Average number of shoots/explant ± (S.E.)*
BAP	1.0	75.0	12.3 ± 0.25
	2.0	80.0	18.0 ± 0.34
	3.0	99.0	29.0 ± 0.45
	4.0	98.0	25.0 ± 0.35
	6.0	73.0	12.0 ± 0.25
	8.0	70.0	10.0 ± 0.35
Kn	1.0	65.0	10.4 ± 0.35
	2.0	75.0	17.7 ± 0.45
	3.0	80.0	25.7 ± 0.35
	4.0	89.0	20.8 ± 0.95
	6.0	72.0	10.0 ± 0.45
	8.0	70.0	8.5 ± 0.45

Direct multiple shoot bud induction from cotyledon explants cultured on various concentrations of Kn (1.0–8.0 mg/l) was observed. High percentage (89) of responding cultures was found at 4.0 mg/l Kn compared to all other concentrations tested. Whereas more number of shoots were regenerated from cotyledon explants at 3.0 mg/l Kn (25.7 ± 0.35 explant) followed by 4.0 mg/l Kn in which profuse rhizogenesis was also observed. At high concentration of Kn less frequency of shoot organogenesis was observed (Table 1). On the whole, the response of BAP as a sole growth regulator was found to be maximum in inducing shoot organogenesis of all the concentrations tested in comparison to Kn.

Influence of auxin-cytokinin combination such as IAA (0.5/1.0 mg/l) + BAP (1.0–8.0 mg/l) and IAA (0.5/1.0 mg/l) + Kn (1.0–8.0 mg/l) in cotyledon explants showed variable response (Tables 2–3). Auxin, 0.5 mg/l IAA was taken in combination with cytokinin, BAP (1.0 to 8.0 mg/l) showed maximum percentage (95%) of responding cultures and high frequency of shoot regeneration (34.3 shoots per explant) at 3.0 mg/l BAP (Plate Ia). As the BAP concentration was increased up to 8.0 mg/l, gradually the induction of shoot organogenesis was found to be reduced. The cotyledon explants when cultured on MS medium fortified with 1.0 mg/l IAA in combination with 1.0–8.0 mg/l BAP showed maximum shoot induction frequency at 3.0 and 4.0 mg/l BAP (14.0 & 13.5 shoots/explant). At 4.0 mg/l BAP in combination with 1.0 mg/l auxin IAA produced roots in addition to multiple shoot induction. The response in inducing the shoot organogenesis was found to be lesser after enhancing the concentration of auxin from 0.5 to 1.0 mg/l (Table 2).

Table 2: Direct shoots formation from cotyledon explants of *S. surattense* cultured on MS medium supplemented with different concentrations of IAA + BAP

Concentration of growth regulators (mg/l)	*% of cultures responding*	*Average number of shoots/explant ± (S.E.)*
IAA + BAP		
0.5 + 1.0	80	13.3 ± 0.43
0.5 + 2.0	80	23.2 ± 0.43
0.5 + 3.0	95	34.3 ± 0.45
0.5 + 4.0	82	24.0 ± 0.32
0.5 + 6.0	82.3	13.0 ± 0.33
0.5 + 8.0	72	10.0 ± 0.32
1.0 + 1.0	72.5	10.0 ± 0.35
1.0 + 2.0	72	12.3 ± 0.43
1.0 + 3.0	75	14.0 ± 0.32
1.0 + 4.0	82	13.5 ± 0.43
1.0 + 6.0	80	8.0 ± 0.35
1.0 + 8.0	88	5.0 ± 0.32

Similarly, cotyledon explants were cultured on MS medium supplemented with IAA (0.5/1.0 mg/l) in combination with various concentrations of Kn (1.0–8.0 mg/l) (Table 3). The results showed the direct shoot regeneration in all the concentrations and combinations tested. Cotyledons cultured on 0.5 mg/l IAA in combination with different levels of Kn showed highest number of shoots (23.8 ± 0.35) at 3.0 mg/l Kn and more percentage (98) of responding cultures. But at high level of Kn shoot bud induction was reduced along with the response too. When the concentration of IAA was

Table 3: Direct shoot bud proliferation from cotyledon explants of *S. surattense* cultured on MS medium supplemented with different concentrations of IAA + Kinetin

Concentration of growth regulators (mg/l)	*% of cultures responding*	*Average number of shoots/explant ± (S.E.)*
IAA + Kn		
0.5 + 1.0	90	12.5 ± 0.35
0.5 + 2.0	92	18.0 ± 0.43
0.5 + 3.0	98	23.8 ± 0.35
0.5 + 4.0	85	18.2 ± 0.32
0.5 + 6.0	80	13.0 ± 0.38
0.5 + 8.0	75	8.5 ± 0.23
1.0 + 1.0	95	10.2 ± 0.45
1.0 + 2.0	90	10.3 ± 0.35
1.0 + 3.0	97	13.0 ± 0.25
1.0 + 4.0	98	12.0 ± 0.32
1.0 + 6.0	85	7.8 ± 0.25
1.0 + 8.0	80	3.9 ± 0.28

increased to 1.0 mg/l, the induction frequency of shoot organogenesis was gradually decreased. From the results it is clear that BAP showed superiority over Kn when it was used individually or along with the two different concentrations of auxin for inducing direct multiple shoot regeneration.

Microshoots (3–4 cm) developed from cotyledon explants were excised and cultured on ½ strength MS medium supplemented with 0.5 mg/l and 1.0 mg/l IAA (Plate Ib). Profuse rhizogenesis was observed on 0.5 mg/l IAA compared to 1.0 mg/l IAA. 80% of the plants were with well developed roots (5–10) on MS medium containing 0.5 mg/l IAA whereas 50% of plants produced roots with 2–3 roots on 1.0 mg/l IAA.

The plantlets were taken out and transferred to plastic pots containing sterile vermiculite and perlite mixture (1:1) for acclimatization. These acclimatized plants were shifted to pots containing garden soil and maintained in the green house. The survival percentage was found to be 70% and the plants were normal consisting of similar morphological features with the parental plants (Plate Ic).

Discussion

The cotyledon explants cultured on MS medium fortified with different concentrations of cytokinins BAP/Kn alone and in combination with 0.5–1.0 mg/l IAA showed the direct shoot regeneration. Maximum numbers of shoots per explant were formed at 3.0 mg/l BAP/Kn when added to the MS medium as a sole growth regulator. When 0.5 mg/l or 1.0 mg/l IAA added to the MS medium containing BAP/Kn showed the enhanced shoot bud proliferation in cotyledon explants. But the induction of shoot bud proliferation was decreased when the concentration of auxin IAA was increased to 1.0 mg/l in all the explants and concentrations of BAP/Kn tested. High frequency of shoots was induced on MS medium supplemented with 0.5 mg/l IAA + 3.0 mg/l BAP during the present investigations. Thus, from the results it is evident that BAP showed superiority over Kn in inducing organogenesis directly from the cotyledon explants in *S. surattense*. Of the cytokinins used BAP proved as most effective than Kn in inducting shoots. The same findings were recorded in *Capsicum* sp. (Philips and Hubstenberger, 1985).

The cotyledon explants cultured on MS medium supplemented with IAA + BAP combination was found to be more effective in inducing maximum number of shoots in *Capsicum* spp. (Gunay and Rao, 1978; Fari and Czako, 1981). Similarly, it was also reported in *Solanum melongena* (Sharma and Rajam, 1995) and *Capsicum annuum* (Christopher and Rajam, 1996) as it was found during the present investigations. The combination of IAA + BAP showed superiority over IAA + Kn in the present studies. Similar findings were also made in *Petunia* (Rao and Harada, 1974), Tomato (Kartha

Multiple shoots induction on
MS + 0.5 mg/l IAA + 3.0mg/l BAP from
cotyledon after 6 weeks of culture

Rhizogenesis from shoot on ½ strength
MS + 1.0 mg/l IAA after 2 weeks of
culture

In vitro regenerated plant

PLATE I

a–c: Cotyledon culture in *Solanum surathense*

et al., 1976), *Physalis* (Bapat and Rao, 1977), *Solanum melongena* (Sadanandam and Farooqui, 1991) and in two species of niger (Jadimathi *et al.*, 1998). All these reports are in conformity with our present work. From the foregoing it is clear that BAP showed superiority in inducing high frequency multiple shoots over Kn.

Thus, the reproducible regeneration protocols developed in the cotyledon explants during the present investigation can be used for mass propagation of the species and also for genetic manipulation studies to introduce agronomically important traits in *S. surattense.*

Acknowledgement

NRS is grateful to TWAS (Italy) & UNESCO (France) for financial assistance.

References

Bapat, U.A. and P.S. Rao.1977. Experimental control of growth and differentiation in organ cultures of *Physalis minima* Linn. *Z. Pflanzen. Physiol.*, **85**:403–416.

Christopher, T. and M.V. Rajam. 1996. Effect of genotype, explant and medium on *in vitro* regeneration of red pepper. *Plant Cell Tiss. Org. Cult.*, **46**:245–250.

Desai, H.V. and A.R. Mehta.1990. Organogenesis in cultured leaf discs of *Passiflora* sp. In: *Hand book of Plant Tissue & Cell Culture*, pp. 28–30 (eds.) A.R. Mehta and P.N. Bhatt, Acad Book Centre, Ahmedabad.

Fari, M. and M. Czako. 1981. Relationship between position and morphogenetic response of pepper hypocotyl explants cultured *in vitro. Sci. Horti.*, **5**:207–213.

Gamble, J.S. 1986. *Flora of the presidency of Madras*, vol. II, Bishen Singh Mahendrapal Singh, Dehradun.

Gunay, A.L. and P.S. Rao. 1978. *In vitro* plant regeneration from hypocotyl and cotyledon explants of red pepper (*Capsicum*). *Plant Sci. Lett.*, **11**:365–372.

Jadimathi, V.G., H.N. Murthy., H.G. Ashok Kumar. and B. Ravi Shankar. 1998. Plant regeneration from leaf cultures of *Guizotia abyssinica* (Niger) and *Guizotia scabra. Phytomorphology*, **48**:131–135.

Jas Rai, V.T., Y. Mudgil., A. Remakanthan. and V.R. Kannan. 1999. Direct shoot regeneration from cultured leaves of *Passiflora caerulea* L. and field performance of regenerated plants. *Phytomorphology*, **49**:289–293.

Jeong, J.H., H.N. Murthy and K.Y. Pack. 2001. High frequency adventitious shoot induction and plant regeneration from leaves of statice. *Plant Cell Tiss. Org. Cult.*, **65**:123–128.

Kartha, K.K., O.L. Gamborg, J.P. Shyluk and F. Constable. 1976. Morphogenetic investigations on *in vitro* leaf culture of tomato (*Lycopersicon esculentum* Mill. c.v. Starfire) and high frequency plant regeneration. Z. *Pflanzen Physiol.*, **77**:292–301.

Kunjumon, A., V.R. Kannan and V.T. Jasrai. 1996. Plant regeneration from leaf explants of three grain legumes on the same medium. *J. Plant Biochem. Biotech.*, **5**:27–29.

Larkin, P.J. and W.R. Scrowcroft. 1981. Somaclonal variation — A novel source of variability from cell cultures for plant improvement. *Theor. Appl. Genet.*, **60**:197–214.

Lee, M. and R.I. Phillips. 1988. The chromosomal basis of somaclonal variation. *Ann. Rev. Plant Physiol. Plant Mol. Biol.*, **39**:413–437.

Leflay, C., E. Chevreau and L.M. Raboin. 1999. Adventitious shoot regeneration from *in vitro* leaves of several pear cultivars (*Pyrus communis* L.). *Plant Cell Tiss. Org. Cult.*, **25**:99–105.

Murashige, T. and F. Skoog. 1962. A revised medium for rapid growth and bioassay with tobacco tissue culture. *Physiol. Plant.*, **159**:473–497.

Nadkarni, K. 1954. *M. Nadkarni's Indian Materia Medica* (India: Popular Book Depot).

Nayar, M.P., K. Ramamurthy and V.S. Agarwal. 1989. *Economic plants of India*, Vol. I to V, Published by Director B.S.I., Calcutta.

Neeta, D.S., H. Singh, S. Tivarekar and S. Eapen. 2001. Plant regeneration from different explants of neem. *Plant Cell Tiss. Org. Cult.*, **65**:159–162.

Philips, G.C. and J.F. Hubstenberger. 1985. Organogenesis in pepper tissue culture. *Plant Cell Tissue Org. Cult.*, **4**:262–269.

Prasad, R.N. and H.C. Chaturvedi. 1978. *In vitro* induction of shoots and formation of plantlets from segments of leaf, stem and root of *Solanum xanthocarpum* Schrad. & Wendl. *Indian J. Exptl. Biol.*, **16**:1121–1122.

Prasad, R.N., M. Sharma, A.K. Sharma and H.C. Chaturvedi. 1998. Androgenic stable somaclonal variant of *Solanum surattense* Burm. f. *Indian J. Exptl. Biol.*, pp. 1097–1012.

Rajani, M.S. and H.G.V.G. Urs. 1998. Multiple shoot induction from cotyledonary explants of *Tabebuia* spp. *in vitro. Phytomorphology,* **48**:207–213.

Rama Swamy, N., T. Ugandhar, M. Praveen, M. Rambabu and M. Upender. 2005. Induction of Streptomycin-resistant plantlets in *Solanum surattense* through *in vitro* mutagenesis. *Plant Cell Tiss. Organ. Cult.*, **80**:201–207.

Rao, P.S. and H. Harada. 1974. Hormonal regulation of morphognesis in organ culture of *Petunia inflata, Antirhinum majus* and *Pharbitis nil. In*: *Plant Growth Substances*, Hirokwa Tokyo, pp. 1113–1120.

Sadanandam, A. and M.A. Farooqui. 1991. Induction and selection of lincomycin-resistant plants in *Solanum melongena* L. *Plant Sci.*, **79**:237–239.

Sharma, P. and M.V. Rajam. 1995. Genotype, explant and position effects on organogenesis and somatic embryogenesis in egg plant (*Solanum melongena* L.). *J. Expt. Bot.*, **46**:135–141.

Sivarajan, V.V. and I. Balachandran. 1999. *Ayurvedic drugs and their plant sources*, Oxford & IBH Publishing Co. Pvt. Ltd., 66, Janpath, New Delhi.

Subhash, K. and T. Christopher. 1988. Direct plantlet formation in cotyledon culture of *Capsicum frutescens. Curr. Sci.*, **57**:99–100.

Swartz, H.J., R. Bors, F. Mohamed and S.K. Naess. 1990. The effect of *in vitro* pretreatment on subsequent shoot organogenesis from excised *Rubus* and *Malus* leaves. *Plant Cell Tiss. Org. Cult.*, **21**:179–184.

Tissue Culture Studies in Asclepiadaceae — A Review

V. Aruna, C. Kiranmai, S. Karuppusamy, S. Anitha and *T. Pullaiah**

Abbreviations: 2,4-D — 2,4-Dichlorophenoxyacetic Acid, 2iP — 6-γ,γ-dimethylallylaminopurine, AA — Ascorbic acid, AC — Activated charcoal, $AgNO_3$ — Silver nitrate, B_5 — Gamborg medium, BA — 6-benzyladenine, CA — Citric acid, CM — Coconut milk, GA_3 — Gibberellic Acid, IAA — Indole-3-Acetic Acid, IBA — Indole-3-Butyric Acid, KN — Kinetin (6-furfurylamonopurine), MS — Murashige and Skoog medium, NAA — α-Naphthaleneacetic Acid, NN — Nitsch and Nitsch medium, PAA — Phenyl acetic acid, PG — Phloroglucinol, PVP — Poly vinyl pyrrolidone, TDZ — Thidiazuron .

Plant tissue and cell culture technologies are saviors in channelizing the resources of the nature for the benefit of the mankind by conservation of elite and endangered plants, eco-friendly production of drugs and drug intermediates. Cryopreservation of germplasm would help in maintaining the genetic diversity of the endangered population and improved cell and tissue culture technologies would help in producing the active compounds *in vitro* with better productivities without cutting down the natural resources. Tissue culture is useful for multiplying and conserving the species, which are difficult to regenerate by conventional methods and save them from extinction.

Tissue culture studies are meagre for important family Asclepiadaceae. The family comprises a number of endemic and endangered taxa. Most of the *in vitro* studies are confined to a few genera of economically and pharmaceutically important such as *Asclepias, Calotropis, Decalepis, Hemidesmus, Tylophora* etc. One of the main impediments in tissue culture studies in this family is the presence of latex.

* Department of Botany and Biotechnology, Sri Krishnadevaraya University, Anantapur-515003, A.P.

The first attempt in tissue culture of Asclepiadaceae is that of Snyder (1955). He cultured stem pith cells of *Cryptostegia grandiflora* on Knop's solution supplemented with different complex growth factors like coconut milk, casein hydrolysate and IAA. The main objective of the study was to understand the growth and development of latex tubes in culture tissues but it was observed only copious proliferation of parenchymatous cells from explants on medium with coconut milk. The reason may be the imbalance of nutrient medium and plant growth regulators.

Latex producing plants contain considerable amount of lower molecular weight hydrocarbons that could be used as a substitute for fossil liquid fuel. So to meet the needs and to avoid over exploitation Tideman and Hawker (1982, 1983) cultured various explants of *Araujia sericifera*. Plantlets were produced from nodal explants, but more successfully from young leaf slices, both required a high cytokinin and low auxin medium for shoot production. The regenerated plants were easily rooted on ½ MS + 0.5µM NAA and successfully transfered to soil in the glass house. Progress has been made towards the establishment of suspension cultures and somatic embryogenesis in the same species from different explants (Torne *et al.*, 1992, 1997). Torne *et al.* (1992) regenerated plantlets from somatic embryos when cultured immature seeds for ornamental purpose. Later Torne *et al.* (1996) cultured petals of immature flower and observed the effect of photoperiodic control and polyamine content of somatic embryogenesis in *Araujia sericifera*. The highest rate of embryogenesis with a high light intensity (90–100 µmol. $m^{-2} s^{-1}$), low doses of 6 -benzyl adenine (BA) and silver thiosulphate free medium has been reported by Torne *et al.* (1997) with corollas of immature flowers.

Another important ornamental genus *Asclepias* was worked out by many researchers. Mante *et al.* (1989) reported shoot regeneration from cotyledons of different species of *Asclepias*. Clonal propagation of medicinal milkweed, *Asclepias curassavica* from nodal segments was reported by Pramanik and Datta (1986). They observed chromosomal variation in the roots of *in vitro* regenerated plants with higher ploidy, few aneuploids and normal diploids. *In vitro* mass propagation of *Asclepias curassavica* have been successfully established on MS and B_5 media with different auxin-cytokinin combination by Dasgupta *et al.* (1986). Dasgupta *et al.* found genetic variability in callus cultures and also observed variation in morphogenetic responses of regenerants. Karpagam and Rao (2003) carried out micropropagation and organogenesis of *Asclepias curassavica* by using shoot tip, node and immature leaf explants and regenerated plants are fertile, phenotypically similar and smaller compared to *in vivo* plants.

Lee *et al.* (1982) cultured apical shoot tips of *Asclepias erosa* on modified MS medium containing 0.54mM NAA + 44.8mM BA to obtain greatest number of shoots. Shoot tips and nodes of *Asclepias erosa* cultured on modified MS medium containing 0.54mM NAA+ 44.4mM BA and 264 or 492 mM IBA was used for rooting and the rooted cultures were established successfully in soil by Lee and Thomas (1985).

The propagation process from fragmented shoot apices of *Asclepias rotundifolia* were cultured on MS + 10mM BA for shoot development, ½ MS + 0.5mM NAA for rooting was reported by Tideman and Hawker (1982). They also cultured nodes but produced very few shoots. Morphogenic investigations of different explants of *Asclepias rotundifolia* were carried out by Tideman and Hawker (1983).

Excellent research work has been focused on callus and suspension cultures but their objectives were different, such as laticifer development in *Asclepias syriaca* studied by Wilson and Mahlberg (1977), Biesboer (1983) and Dunbar *et al.* (1986). Embryoids and induced shoots of *Asclepias syriaca* were established from stem explants in which it was possible to control for the first time the differentiation of laticifer as a cell type in structure similar to that present in the natural plant (Wilson and Mahlberg, 1977). Dunbar *et al.* (1986) identified laticifers in frozen sections of embryoids from callus and suspension cultures of *Asclepias syriaca* by an indirect fluorescent antibody technique. Vangala *et al.* (1985) compared protoplast yield from leaves, callus culture and suspension culture of *Asclepias syriaca*. Vangala *et al.* (1985) reported 2.4×10^5 protoplast/ml with leaf mesophyll, $3.0–3.3 \times 10^5$ protoplast/ml with callus tissue and $1.0–2.0 \times 10^5$ protoplast/ml with suspension culture and reported that callus tissue was the best source of protoplast because of its yield and viable cells. Knorr and Miazga (1987) carried out batch cultures of freely suspended as well as calcium alginate gel or chitosan gel entrapped cells of *Asclepias syriaca* and detected proteolytic activity in the cell homogenates and in the culture media from freely suspended and gel entrapped milkweed cells. Different physiological changes during suspension cultures in bioreactors, O_2 uptake, change in pH, cell viability, packed cell volume were studied by Biesboer and Tate (1988). Tepper and Knapp (1992) excised cotyledons from mature embryos of *Asclepias syriaca* and cultured on MS medium containing 2.5mM IBA + 7.5mM BA and produced adventitious shoots. They also reported that cells of proximal, adaxial region were competent to perform regenerative function.

Extensive morphogenetic and suspension culture studies in *Asclepias tuberosa* were carried out by various researchers. Suspension culture and somatic embryogenesis were reported by Groet and Kidd (1980), Biesboer (1984) and Wilson and Dunbar (1987). Vangala *et al.* (1985) studied the protoplast yield whereas Dunbar *et al.* (1986) studied the laticifer cell development in callus and embryoids. Different physiological changes during suspension cultures in bioreactors were studied by Biesboer and Tate (1988). By using nodal segments as explants Vyapari *et al.* (1993) produced plantlets and successfully transplanted on to soil.

Researchers have shown interest in latex producing plants containing considerable quantities of low molecular weight hydrocarbon that could be used as a substitute for fossil fuels. In the list of petrocorps *Calotropis gigantea* has been included due to its latex containing long chain hydrocarbons equivalent to crude oil. So *in vitro* propagation and enhancement of hydrocarbons in this plant is quite needed. The first attempt in tissue culture of *Calotropis gigantea* is that of Datta (1981). He cultured shoot tips on modified MS medium with auxins and obtained callus from which rhizogenesis was observed. He also attempted to induce shoot regeneration from callus derived from shoot tips, but he was unsuccessful. De Bagga *et al.* (1986) and Dutta and De (1986) worked for understanding laticifer cell differentiation from callus cultures of *Calotropis gigantea* and also determined secondary metabolites from *in vitro* cultured tissue such as cardenolides, where as Pramanik *et al.* (1986) estimated cardioglycoside from *in vitro* cultured tissue. At the same time researchers also focused on morphogenesis (De Bagga *et al.*, 1986; Pramanik *et al.*, 1986; Roy and De, 1987, 1988, 1990). Stem and hypocotyl of *Calotropis gigantea* were grown on different concentration of auxins by Roy and De (1987). Then they transferred organogenic callus to 4.44mM BA and 0.54mM NAA and produced 8–10 shoot buds per callus and microshoots were rooted with Bonner's solution. Roy and De (1988) established callus culture from immature embryos, hypocotyl, stem, leaf tissue of *Calotropis gigantea* on modified MS medium with myo-inositol 300mg/l + glycine 3mg/l + casein acid hydrolysate 600mg/l and obtained high frequency of multiple buds and shoots which was subsequently rooted in Bonner's salt solution with IAA at low concentration. De (1989) cultured callus of *Calotropis gigantea* in modified MS medium supplemented with 10.74mM NAA and produced maximum rhizogenesis and reported that amount of protein synthesized during rhizogenesis was more than that of DNA and RNA but prior to the formation of roots, large amount of RNA and proteins are synthesized and stimulated rhizogenesis. In the same year Roy and De (1989, 1991) reported somatic embryogenesis of *Calotropis gigantea* from hypocotyl explants. Immature embryos of *Calotropis gigantea* were cultured by Roy and De (1990) on modified MS medium + 4.52mM 2,4-D, and shoot bud formation from callus was reported on NAA 0.54mM +BA 8.87mM. Rooting was induced in auxin supplemented Bonner's solution or half strength MS basal salt solution. Roy and De (2000) obtained friable callus by culturing the hypocotyl explant of *Calotropis gigantea* on 2.69μM NAA + 4.44 μM BA. Shoots were regenerated on 2.22μM BA + 0.45μM 2, 4-D and rooted on Bonner's solution containing 0.49μM IBA.

A few reports are available on the other related species *Calotropis procera*. Stem explants of *Calotropis procera* were cultured on MS medium + 6.78mM 2,4-D + 2.32mM KN + PVP by Dhir *et al.* (1984). Dhir *et al.* produced callus and reported that laticifer cells were not present in young callus but were observed after 4 weeks of callus growth. Positive Liebermann Burchard test proved the presence of terpenoids in laticifers. Later Dhir and Maheshwari (1986) developed a protocol of high yield preparation of viable protoplast from leaves of *Calotropis procera*. Laticifer studies of *Calotropis procera* were further carried out by Suri and Ramawat (1995, 1997). Suri and Ramawat (1995) cultured cotyledons on MS medium + 4.6mM KN + 1µM IAA resulted with maximum laticifer differentiation (17.01%). Suri and Ramawat (1996) studied the effect of *Calotropis* latex on laticifer differentiation in callus culture of *Calotropis procera*. *In vitro* propagation of *Caralluma edulis* was reported by Kaur *et al.* (1992).

Ceropegia is one of the endangered taxon in the Indian subcontinent. Ananthan (2003) carried out micropropagation and conserved *Ceropegia barnesii* and *Ceropegia pusilla*. Patil (1998) reported micropropagation and somatic embryogenesis in three different taxa such as *Ceropegia bulbosa*, *Ceropegia bulbosa* var. *lushii* and *Ceropegia jainii*. He cultured nodal segments of three species and reported that ½ MS + 8.8µM BA gave best response. Rhizogenesis was reported on ¼ MS + 2.4µM IBA. He also observed microtuber formation with 46.5µM KN + 22.2µM BA. Stem derived callus of all three *Ceropegia* genotypes when transferred to 0.5 MS + 2.2µM BA developed somatic embryos with bipolar structure which were further transferred on to regeneration medium, for development into full plantlets. Patil reported *in vitro* flowering in *Ceropegia jainii* on MS + 22.2µM BA for the first time in the family Asclepiadaceae. While producing aseptic seedling in *Ceropegia bulbosa* for *in vitro* studies Raghuramulu *et al.* (1999) reported twin seedlings. Raghuramulu and Pullaiah (1999) reported multiple shoots with root, cotyledon, leaf, stem segments and plantlets were regenerated and rooted *in vitro* followed by acclimatization. A rapid micropropagation system was developed for *Ceropegia bulbosa* by John Britto *et al.* (2003). Nodal explants cultured on B_5 + BA 13.3mM + NAA 0.269mM produced a maximum of up to 12 shoots with in one week. They also reported *in vitro* flowering on BA 2.22mM + GA_3 2.89mM (76%) leading to seed production and 65% of the seeds germinated in successive generations. Beena *et al.* (2003) cultured axillary bud of *Ceropegia candelabrum* on MS + 8.87µM BA + 2.6µM IBA and produced 8 shoots/node. Shoots developed were rooted best on ½ MS+0.49µM IBA. Starting from a single nodal explant 205 rooted shoots were obtained within 120 days by Beena *et al.* At the same time Beena and Martin (2003) cultured leaf and internodal explants on MS + 4.52µM 2, 4-D and obtained friable callus, which underwent somatic embryogenesis. Friable callus was transferred to suspension culture on half or quarter strength MS medium with lower levels of 2,4-D (0.23 or 0.45 μM) and induced the highest number of somatic embryos. About 100 mg of callus induced more than 500 embryos among which 50% of somatic embryos underwent maturation and plantlets were acclimatized under field condition with 90% survival. In four weeks 10–12 shoots were developed per axillary bud on BA 8.87mM of *Ceropegia noorjahani* by Kedge *et al.* (2006). The shoots were rooted on ½ MS + IBA 9.8mM + NAA 5.37mM + 200mg/l activated charcoal. Direct plant regeneration was obtained from 100 % of the nodal explant of *Ceropegia oculata* on MS + 7.5μM BA + 3% sucrose by Nikam and Savant (2006). Regenerated shoots were rooted on liquid ½ MS fortified with 5% sucrose + 6.0mg/l spermine.

Milkweeds are known to contain cytotoxic glycosides. Venkateswara *et al.* (1987) isolated cryptosin a new steroidal glycoside, which is found to be cardioactive, from cultured tissue of *Cryptolepis buchanani.* Metabolically active callus of *Cryptolepis* were proliferated and maintained on B_5 medium with 2,4-D and KN. Prasad *et al.* (2004) obtained 12.5 to 13.0 shoots/explant on MS + BA 8.87mM + KN 0.46mM + NAA 0.266mM + GA_3 0.115mM from nodal explant in *Cryptolepis buchanani* and shoots are rooted on MS + IBA 4.9mM.

Raghuramulu (2001) reported direct shoot regeneration from root explant of *Cynanchum callialatum*. Besides this he also reported shoot regeneration from callus from different explants and somatic embryogenesis from immature embryos. A rapid *in vitro* propagation by nodal explant of *Cynanchum callialatum* was carried out by Raghuramulu *et al.* (2002). They reported highest number of shoots (3.05shoots) with BA 4.44mM + KN 2.32mM + GA_3 1.44mM + 50mg/l AA + 10% CM. The regenerated shoots were rooted on ½ MS + IAA 5.71mM + KN 0.93mM and 85% of the transfered plants survived. Haccius and Hausner (1976) worked on unpollinated ovaries *of Cynanchum nigrum* and *Cynanchum vincetoxicum.* Shin *et al.* (2003) attempted to synthesize gagaminine which is a steroidal alkaloid from *Cynanchum wilfordii*. Callus was induced from the stem on MS + 2, 4-D 9.05mM and gagaminine production was increased markedly when cell growth proceed to the death phase.

Decalepis arayalpathra, is a rare endemic medicinal plant. Fast growing normal root culture was established from leaf and internodal explants on MS + 11.09mM BA + 2.46mM 2iP + 2.69mM NAA by Sudha and Seeni (2001). Leaf explants incubated in darkness in ½ MS + 2.46mM IBA + 1.07mM NAA favoured induction of roots with highest biomass and reported that 100mg fresh root tissue cultured in 80ml ½ MS liquid medium + 0.98mM IBA + 0.54mM NAA yielded 3.433g and 0.734g fresh and dry weight of root. Gangaprasad *et al.* (2005) cultured single node from top shoots of *Decalepis arayalpathra* in MS + BA 4.44mM and produced the longest shoots. The shoots were rooted on 8.65mM IAA. Sudha *et al.* (2005)

cultured basal nodes of 12–16 week old *Decalepis arayalpathra* in MS + 12.96μM BA +2.48μM 2iP + 268 μM NAA and produced 16–17 cm long unbranched robust solitary shoots in 8 weeks. Rooting was carried out on ½ MS + 1.07μM NAA.

Decalepis hamiltonii root is well known for its medicinal value. Bais *et al.* (2000a) cultured axillary bud of *Decalepis hamiltonii* and produced maximum number of 12.8 shoots on MS + BA 8.87mM + NAA 2.69mM. Bais *et al.* (2000b) studied the effect of silver nitrate on root formation of *Decalepis hamiltonii.* Treatment of 40μM $AgNO_3$ promoted maximum rooting of microshoots. The promotive effects of $AgNO_3$ on rooting may be from inhibition of ethylene action. George *et al.* (2000) succeeded in regenerating shoots from leaf explants of *Decalepis hamiltonii.* Obul Reddy *et al.* (2001) reported that cobalt chloride, charcoal and phloroglucinol (PG) along with IBA treatment had beneficial effect on rooting of *in vitro* derived shoots of *Decalepis hamiltonii. In vitro* multiple shoot formation from nodal explant was reported in *Decalepis hamiltonii* by Anitha and Pullaiah (2002). Anitha and Pullaiah produced highest number of 10–14 shoots on MS + 13.3mM BA + 2.32mM KN + 2.69mM NAA + 10% CM +150mg/l AA. Anitha (2002) succeed in the proliferation of callus from different explants of *Decalepis hamiltonii,* but failed in regenerating plantlets. Obul Reddy *et al.* (2002) studied the effect of Triacontanol on micropropagation of *Decalepis hamiltonii.* Giridhar *et al.* (2003) reported a protocol.for shoot multiplication and elongation of *Decalepis hamiltonii* using Phenyl acetic acid (PAA) in the medium. The presence of 31.08μM BA in combination with 14.68μM PAA was used for the induction of primary shoots from nodal explant and further regenerated shoots were rooted on 9.8 μM IBA. Axillary buds of *Decalepis hamiltonii* were used to initiate multiple shoots on MS + 8.87mM BA + 2.85mM IAA by Giridhar *et al.* (2004). The shoots were rooted on MS + 1.0mg/l IBA and regenerated plants were grown successfully in plains. A maximum of 0.14% and 0.12% 2-hydroxy-4-methoxybenzaldehyde was produced in roots of one year old tissue culture derived plants and greenhouse grown plants respectively. A highly efficient two stage protocol was developed for multiple shoots from single node by Gururaj *et al.* (2004) in *Decalepis hamiltonii.* Maximum number of shoots per culture was reported on the medium containing 1.1μM BA + 5.8μM GA_3 + 200μM PG+ 0.011μM Triacontanol for shoot elongation. The long shoots were reported on 5.38μM NAA+ 400μM PG and plantlets were hardened with 80–90% survival rate. Giridhar *et al.* (2004) produced plantlets of *Decalepis hamiltonii* from leaf explants via somatic embryogenesis and organogenesis. An efficient and rapid micropropagation system was developed for *Decalepis hamiltonii* by Giridhar *et al.* (2005) using shoot tip explant.

Komalavalli and Rao (1997) cultured seedling and mature axillary nodes of *Gymnema elegans* and produced maximum number of shoots on

MS +13μM BA + 2μM KN + 3μM NAA + 100mg/l AA + 10%CM + 1μM GA_3. Seedling derived shoots were rooted on MS+ 5μM IBA whereas, shoots derived from mature explants were rooted on MS + 15μM IBA and developed shoots were hardened successfully.

Another related antidiabetic plant *Gymnema sylvestre* was propagated from nodal segments by Anu *et al.* (1994). The combination of low cytokinin and relatively high IAA favoured bud break and later growth of shoot. Addition of 30% CM along with 2ppm adenine sulphate slightly enhanced sprouting. Sairam Reddy *et al.* (1998) developed a rapid system for regenerating shoots from mature nodal explant of *Gymnema sylvestre*. Maximum number of 7 shoots were produced on BA 22.19mM + NAA 1.07mM and regenerated shoots were rooted on ½ MS without growth regulators. Different seedling explants like node, leaf, shoot tip, cotyledonary node, petiole, root, internodal segments were cultured and produced maximum number of 57.2 shoots with nodal explant on MS + 4.44mM BA + 2.32mM KN + 0.54mM NAA + 100mg/l malt extract + 100mg/l CA by Komalavalli and Rao (2000). High frequency of rooting was reported on ½ MS + 14.7mM IBA. Somatic embryogenesis and whole plant regeneration was recorded from hypocotyl, cotyledon and leaf explant of *Gymnema sylvestre* by Ashok Kumar *et al.* (2002). They induced embryogenic callus on MS + 2,4-D (0.5–5.0μM) + BA (0.5–2.0μM) and then subcultured on to embryo maturation medium containing MS salts, B_5 vitamins, 0.5μM BA and 2% sucrose.

The Indian sarsaparilla *Hemidesmus indicus* was propagated through *in vitro* methods by Heble and Chadha (1978), Sarasan *et al.* (1994), Sharma and Yelne (1995), Patnaik and Debata (1996), Malathy and Pai (1998), Sree Kumar *et al.* (2000), Raghuramulu (2001), Saha *et al.* (2003), Neeta *et al.* (2003). Sarasan *et al.* (1994) induced callus from leaf and stem explants of *Hemidesmus indicus*. Callus induced on MS + 2, 4-D 4.52mM + KN 2.32mM developed somatic embryos. Organogenesis was induced on NAA 10.74mM + KN 2.32mM and subcultured to 1.5–2mg/l KN and 10% CM. The shoots were rooted on ½ MS basal medium. Patnaik and Debata (1996) cultured nodal explants of *Hemidesmus indicus* on MS + 1.15 μM KN + 0.054μM NAA. The shoots were rooted on 1.15 μM Kinetin + 7.35μM IBA. Whereas Malathy and Pai (1998) reported a maximum of 13 shoots on MS medium +BA 11.09mM + KN 2.32mM. Sree Kumar *et al.* (2000) used nodal explant of *Hemidesmus indicus* for improved concentration of 2-hydroxy-4-methoxybenzaldehyde in the micropropagted plants. The root biomass of micropropagated plant is 13.5g compared to the plant raised from conventional rooted (4.5g) stem cutting. Shoot and callus cultures from root and leaves of *Hemidesmus indicus* was established by Neeta *et al.* (2005). They produced antioxidants such as lupeol, vanillin and rutin from root and shoot cultures. An *in vitro* protocol for plant regeneration from callus via somatic embryogenesis has been developed in *Hemidesmus indicus* using leaf explant by Swaroopa and Dixit (2006). They obtained compact yellowish embryogenic callus on 2, 4-D 4.52mM which resulted in 10–12 somatic embryos on NN basal medium in dark condition.

Genetic transformation and hairy root induction in *Holostemma ada-kodien* was carried out by Karmarkar and Keshava Chandran (2001). A protocol was described by Martin (2002) for the micropropagation of *Holostemma ada-kodien* through axillary bud multiplication and indirect organogenesis. He got 15 shoots from basal cut end of nodal explant on MS + 6.66mM BA. Callus from internodal explant developed fewer shoots whereas leaf derived callus did not undergo organogenesis. Microshoots cultured on ½ solid MS or liquid medium with 0.246mM IBA exhibited best *in vitro* rooting. Plant regeneration through indirect somatic embryogenesis was attempted from leaf, internode, and root explants derived callus of *Holostemma ada-kodien* by Martin (2003). An average of 40 embryos were obtained from 10 mg callus with 90% field survival.

Sudha *et al.* (1998) established micropropagation system of *Holostemma annulare* by nodal culture. They induced multiple shoot formation on 4.43µM BA + 0.5µM NAA and rooted on ½ MS +1.48µM IBA. Shoot multiplication and cryopreservation of *Holostemma annulare* was carried out by Decruse *et al.* (1999), Decruse and Seeni (2002), Decruse *et al.* (2004). Shoot tip of *Holostemma annulare* from shoot culture raised on MS-2 (NH_4^+ + NO_3^- - 2.6mM + 18.8mM) upon preconditioning in MS-2 or MS-3 (NH_4^+ + NO_3^- - 0.0mM + 18.8mM) and subsequent preculture of encapsulated shoot tips and post-freeze recovery culture in MS-3 showed maximum regeneration (55%) (Decruse *et al.*, 2004). Sudha *et al.* (2000) regenerated shoots from *in vitro* root segments of *Holostemma annulare.*

Maraffa *et al.* (1981) used leaf explants of *Hoya carnosa* var. *compacta* and *Hoya carnosa* var. *rubra* for proliferation of callus in the presence of 2,4-D and KN. Secondary culture, apparent asexual embryos were developed on NAA + KN.

Indirect somatic embryogenesis and subsequent plant regeneration were established from leaves in *Leptadenia reticulata* by Hariharan *et al.* (2002). MS + 8.87mM BA + 2.69mM NAA was found to be more effective combination for induction of embryogenic callus. Embryoids developed on full/half strength hormone free MS or with lower concentration (1mg/l) of cytokinins (BA or KN). Adventitious roots generated using meristem explants of *Leptadenia reticulata* was reported by Bhatt *et al.* (2002). Meristem explants were cultured on Murashige and Skoog (MS) medium supplemented with IBA (3µM). After 30 days, subculturing of roots into liquid MS medium containing IBA (3 mM) for *Leptadenia reticulata* afforded considerable increase in root mass. HPTLC profiles and microscopic examination of transverse sections of *in vitro* and naturally grown roots provided information on secondary metabolite accumulation. Arya *et al.*

(2003) carried out micropropagation with nodal explant in *Leptadenia reticulata*. Martin (2004) attempted to regenerate plants through indirect somatic embryogenesis from leaf, internode, node and shoot-tip derived callus of *Leptadenia reticulata*. Only shoot-tip and node derived calli induced somatic embryos and then transferred embryogenic callus to suspension cultures. Suspension cultures with reduced concentration of BA (2.22 mM) either alone or in combination with 0.49 mM IBA fostered maturation of embryos. Half-strength MS solid medium with 1.44 mM GA_3 and BA (0.22 or 0.44 mM) facilitated conversion of embryos into plantlets at higher rate compared to that on with BA alone. From 10 mg of callus Martin recovered about 77 plantlets.

Callus tissue derived from stem segments was used for protoplast isolation, regeneration and propagation in *Pergularia pallida* by Bapat *et al.* (1986) and Mhatre *et al.* (1987). Prasad (2004) cultured *Sarcostemma intermedium* for shoot multiplication and for plant regeneration from callus culture from different explants. Maximum number of multiple shoots (5.25 shoots/explant) from the nodal explants of *Sarcostemma intermedium* was observed on MS medium containing BA (11.09mM) + Zeatin (0.0228mM) + NAA (0.054mM) + GA_3 (0.029mM). Callus proliferated on 2, 4-D 4.52mM and Picloram 1.03mM were selected for organogenesis. Maximum shoot induction from callus raised on 2,4-D 4.52mM was observed on MS medium fortified with BA 1.11mM + KN 0.46mM + TDZ 0.0227mM + NAA 0.269mM (1.40 shoots). MS half strength fortified with IBA 2.46mM alone and IBA 2.46mM + NAA 0.269mM in combination are effective for better rooting. Amariei *et al.* (1991) observed shoot regeneration from callus cultures of *Solanostemma argel*.

Several reports have been published on *in vitro* studies of *Tylophora*, an important medicinal plant. Nodal explant of *Tylophora asthmatica* was used for multiple shoot induction by Usha *et al.* (2002). The first report of somatic embryogenesis was noticed in *Tylophora indica* by Rao *et al.* (1970). They repeatedly subcultured callus on to different nutrient media in sequence and regenerated roots, shoots and typical bipolar embryos *in vitro*. The early investigation on callus proliferation and embryogenesis from stem explants was noticed in *Tylophora indica,* a potent medicinal plant, by Rao and Narayanaswamy (1972). Very limited research has been carried out *in vitro* secondary metabolite production from callus cultures. Benjamin and Mulchandani (1973, 1976) are the pioneer researchers in carrying studies on secondary metabolites from callus cultures of *Tylophora indica*. From the same species Benjamin *et al.* (1979) regenerated the shootlets and detected the presence of phenanthroindolizidine alkaloids from callus cultures. Mhatre *et al.* (1984, 1987) and Rao *et al.* (1986) isolated protoplasts from callus derived from stem segment and regenerated plant in *Tylophora indica*.

A procedure for rapid *in vitro* multiplication by axillary bud of *Tylophora indica* was reported by Sharma and Chandel (1992). Optimum multiplication was observed on MS + BA 22.19mM + NAA 2.69mM + 100mg/l AA. *In vitro* shoots of *Tylophora indica* were rooted on IAA 5.71mM. Chattopadhyay *et al.* (1992) used nodal explants of *Tylophora indica* to study the effect of NH_4NO_3 on alkaloid synthesis in regenerated shoots. Numerous shoots were present after 60–70 days of culture in 0.16% NH_4NO_3 but the growth rate was maximum in 0.30% of NH_4NO_3 and also the regenerated plants were rich in alkaloid content. Shoot multiplication by using nodal explant was carried out by Bera and Roy (1993) and Sharma *et al.* (1995). Direct shoot regeneration and plantlet propagation of *Tylophora indica* using leaf explant was reported by Bera and Roy (1993). With the same explant somatic embryogenesis was obtained (Bera and Roy, 1995, Manjula *et al.*, 2000, Jayanthi and Mandal, 2001). Bera and Roy (1995) found that rate of synthesis of protein, DNA, RNA were higher in the embryonic cells than in non embryonic cells during the differentiation of single cell to globular embryos. RAPD analysis of regenerated plants (Jayanthi and Mandal, 2001) was worked out in *Tylophora indica*. Adventitious shoot regeneration from leaf (Faisal and Anis, 2003), stem segments (Faisal and Anis, 2005) and petiole (Faisal *et al.*, 2005) were also reported in *Tylophora indica*. Both leaf and stem segments were cultured on 10mM 2, 4, 5-T. and adventitious shoots were regenerated from the surface of the callus on 5mM Kinetin whereas petiole was cultured on 10mM 2,4-D + 2.5mM TDZ and shoot multiplication was achieved on 2.5mM TDZ. The developed shoots were rooted on 0.5mM IBA. Chaudhuri (2004) reported direct plantlet regeneration *in vitro* from root segments of *Tylophora indica*. Root explants cultured on BA or 2iP produced organogenic nodular meristemoids resulted in shoot bud formation and somatic embryogenesis. Genetic transformation, synthesis and secretion of tylophorine from intact shoots was reported by Chaudhuri *et al.* (2005). An efficient and reproducible protocol for high frequency callus formation from immature leaf pieces was obtained by Thomas and Philip (2005) in *Tylophora indica*. They reported shoot regeneration from 100% cultures with an average of 66.7 shoots per culture.

Emmanuel *et al.* (2001) cultured node, shoot tip of *Wattakaka volubilis* for production of shoots. Whereas Chakaradhar (2004) cultured node, shoot tips, root, leaf, internode for plant regeneration. Callus mediated shoot organogenesis was obtained by sub culturing the organogenic callus obtained from leaf explants on MS + 5.37mM NAA + 2.22mM BA to a fresh medium fortified with 4.44mM BA + 0.29mM GA_3 lacking NAA. Highest number of shoots obtained from nodal explants cultured on 23.23mM KN in combination with 0.54mM NAA.

Table 1: Tissue culture studies in Asclepiadaceae

Species name	*Type of culture*	*Source of Explant*	*Result*	*Reference*
Araujia sericifera (Cruel Plant)	Direct/callus	Transverse leaf slices	Adventitious shoots and rooted *in vitro* embryogenesis	Tideman and Hawker (1982)
Araujia sericifera	Shoot/callus	Stem node	Some callus, axillary shoots, rooted *in vitro*	Tideman and Hawker (1982)
Araujia sericifera	Shoot	Shoot tip, leaf	Shoot regeneration	Tideman and Hawker (1983)
Araujia sericifera	Direct/callus	Immature seeds	Somatic embryos, plantlets	Torne *et al.* (1992)
Araujia sericifera	Callus	Petals from immature flowers	Polyamine and photocontrol of Somatic embryogenesis	Torne *et al.* (1996)
Araujia sericifera	Direct/somatic embryos	Corollas of immature flowers	Somatic embryogenesis, plantlets	Torne *et al.* (1997)
Asclepias curassavica	Shoot	Node	Shoot regeneration with ploidy variation and rooted *in vitro*	Pramanik and Datta (1986)
Asclepias curassavica	Callus	–	Mass propagation and genetic variability	Dasgupta *et al.* (1986)
Asclepias curassavica	Callus	Stem segments	Callus morphogenesis, cardioglycoside estimation	Pramanik *et al.* (1986)
Asclepias curassavica	Callus/ shoot	Shoot tip, leaf, node, immature leaf	Multiple shoots, organogenesis, shoots rooted *in vitro*	Karpagam and Rao (2003)
Asclepias erosa	Shoot	Apical Shoot tip	Multiple shoots, rooted *in vitro*	Lee *et al.* (1982)
Asclepias erosa	Shoot	Shoot tip, stem node	Multiple shoots, rooted *in vitro*	Lee and Thomas (1985)
Asclepias rotundifolia	Shoot	Stem node	Axillary shoot proliferation, shoots rooted *in vitro*	Tideman and Hawker (1982)
Asclepias rotundifolia	Shoot	Fragmented shoot apices	Shoot growth, axillary shoot proliferation, rooted	Tideman and Hawker (1982)
Asclepias rotundifolia	Shoot	Shoot tip	Shoot regeneration	Tideman and Hawker (1983)
Asclepias sp.	Shoot	Cotyledons	Adventitious shoots	Mante *et al.* (1989)
Asclepias syriaca (Common milkweed)	Callus	Stem	Callus culture, embryogenesis, laticifer differentiation	Wilson and Mahlberg (1977)
Asclepias syriaca	Suspension	Callus from zygotic embryos	Embryogenesis, plantlets.	Groet and Kidd (1980)
Asclepias syriaca	Suspension/ callus	Leaf, stem	Callus proliferation, cell suspension culture and detection of laticifer cell development	Biesboer (1983)
Asclepias syriaca	Callus	Mature embryos	Callus from hypocotyls, embryogenesis, plantlets	Singh (1984)

Asclepias syriaca	Protoplast	Hypocotyl pieces	Protoplast, Cell colonies, shoot regeneration	Singh (1984)
Asclepias syriaca	Suspension/ callus	Hypocotyl/callus	Adventitious shoots, rooted plantlets	Singh (1984)
Asclepias syriaca	Protoplast	Leaf, callus and suspension cells from stem	Callus culture, suspension culture, Protoplast yield	Vangala *et al.* (1985)
Asclepias syriaca	Callus	Stem sections of aseptic seedling	Callus culture, somatic embryogenesis, identification of laticifer cells	Dunbar *et al.* (1986)
Asclepias syriaca	Suspension	Hypocotyl, primary leaves of aseptic seedling	Protease production from immobilized cells	Knorr and Miazga (1987)
Asclepias syriaca	Suspension	–	Suspension culture, bioreactors, O_2 uptake, change in pH, cell viability, packed cell volume	Biesboer and Tate (1988)
Asclepias syriaca	Shoot	Cotyledons	Adventitious shoots, rooted plantlets	Tepper and Knapp (1992)
Asclepias tuberosa (Butterfly flowers)	Suspension	Callus from zygotic embryos	Suspension culture, Embryogenesis, plantlets	Groet and Kidd (1980)
Asclepias tuberosa	Suspension	Leaf	Suspension culture, Somatic embryogenesis	Biesboer (1984)
Asclepias tuberosa	Protoplast	Leaf, callus and suspension cells from stem	Callus culture, suspension culture, Protoplast yield	Vangala *et al.* (1985)
Asclepias tuberosa	Callus	Stem sections of aseptic seedling	Callus culture, somatic embryogenesis, no laticifer cells	Dunbar *et al.* (1986)
Asclepias tuberosa	Suspension	Embryoids	Suspension culture, somatic embryogenesis	Wilson and Dunbar (1987)
Asclepias tuberosa	Suspension	–	Suspension culture, bioreactors, O_2 uptake, change in pH, cell viability, packed cell volume	Biesboer and Tate (1988)
Asclepias tuberosa	Callus/Shoot	Node	Callus, Multiple shoot formation, rooted *in vitro*	Vyapari *et al.* (1993)
Calotropis gigantea	Callus	Shoot apices	Callus proliferation, rhizogenesis from callus	Datta (1981)
Calotropis gigantea	Callus	Stem segments	Callus culture, laticifer differentiation, cardenolide	De Bagga *et al.* (1986)
Calotropis gigantea	Callus	Stem segments	Callus culture, morphogenesis, Cardioglycoside estimation	Pramanik *et al.* (1986)
Calotropis gigantea	Callus		Laticifer differentiation	Dutta and De (1986)

contd...

... *contd.* Table 1

Species name	*Type of culture*	*Source of Explant*	*Result*	*Reference*
Calotropis gigantea	Callus	Stem segments, hypocotyl	Callus proliferation, shoot regeneration, propagation of plantlets	Roy and De (1987)
Calotropis gigantea	Callus	Immature embryo hypocotyl, stem, leaf segments	Callus proliferation, adventitious shoot induction, plantlet propagation	Roy and De (1988)
Calotropis gigantea	Callus	Shoot tip	Callus culture, rhizogenesis from callus, DNA, RNA and Protein content during rhizogenesis	De (1989)
Calotropis gigantea	Callus/ suspension	Hypocotyl	Callus culture, suspension culture, somatic embryogenesis	Roy and De (1989)
Calotropis gigantea	Callus	Immature embryos	Callus culture, shoot bud formation, plantlet propagation	Roy and De (1990)
Calotropis gigantea	Callus/ suspension	Hypocotyl	Callus culture, suspension culture, somatic embryogenesis, propagation of plantlet	Roy and De (1991)
Calotropis gigantea	Callus/ suspension	Hypocotyl	Cell suspension culture, plant regeneration	Roy and De (2000)
Calotropis procera	Callus	Node, internode segments of stem	Callus culture, detection of laticifer cell development	Dhir *et al.* (1984)
Calotropis procera	Protoplast	Leaf	Protoplast isolation	Dhir and Maheshwari (1986)
Calotropis procera	Callus	Cotyledons from aseptic seedlings	Callus culture, laticifer differentiation	Suri and Ramawat (1995)
Calotropis procera	Callus	Cotyledons	Callus culture, laticifer differentiation	Suri and Ramawat (1996)
Calotropis procera	Callus	Cotyledons	Callus culture, detection of role of calcium in the laticifer differentiation	Suri and Ramawat (1997)
Calotropis procera	Callus	–	Callus culture, cardiac glycosides	Zafar and Ranju Rana (2004)
Caralluma edulis	Shoot	–	*In vitro* propagation	Kaur *et al.* (1992)
Ceropegia barnesii	Shoot	Different explants	Micropropagation , conservation	Ananthan (2003)
Ceropegia bulbosa	Shoot	Node	Multiple shoot induction, plantlet propagation	Patil (1998)
Ceropegia bulbosa	Callus	Stem segments	Callus culture, plantlet propagation	Patil (1998)
Ceropegia bulbosa	Aseptic seedling	Seeds	Polyembryony	Raghuramulu *et al.* (1999)

Ceropegia bulbosa	Shoot	Node	Multiple shoot induction, plantlet propagation	Raghuramulu and Pullaiah (1999)
Ceropegia bulbosa	Callus	Root, cotyledon, leaf, stem segments	Callus culture, plantlet propagation	Raghuramulu and Pullaiah (1999)
Ceropegia bulbosa	Shoot	Node	Multiple shoot formation, *in vitro* flowering, microtubers	John Britto *et al.* (2003)
Ceropegia bulbosa var. *lushii*	Callus	Stem segments	Callus culture, plant regeneration	Patil (1998)
Ceropegia bulbosa var. *lushii*	Shoot	Node	Multiple shoot induction, plantlet propagation	Patil (1998)
Ceropegia candelabrum	Shoot	Node	Axillary bud multiplication, rooted *in vitro*	Beena *et al.* (2003)
Ceropegia candelabrum	Callus	Leaf, internodal segments	Callus induction, somatic embryogenesis, plantlets from embryoids	Beena and Martin (2003)
Ceropegia jainii	Shoot	Node	Multiple shoot induction, plantlet propagation	Patil (1998)
Ceropegia jainii	Callus	Stem segments	Callus culture, somatic embryogenesis	Patil (1998)
Ceropegia noorjahani	Shoot	Axillary bud	Multiple shoot formation, rooted *in vitro*	Kedage *et al.* (2006)
Ceropegia oculata	Shoot	Node	Shoot multiplication, rooted *in vitro*	Nikam and Savant (2006)
Ceropegia pusilla	Shoot	Different explants	Propagation, conservation	Ananthan (2003)
Cryptolepis buchnani	Callus	Leaf, internodal segments	Callus culture, determination of cryptosin concentration	Venkateswara *et al.* (1987)
Cryptolepis buchanani	Shoot	Node from aseptic seedling	Multiple shoots, rooted *in vitro*	Prasad *et al.* (2004)
Cryptolepis buchanani	Callus	Leaf, cotyledon, internode, root	Callus culture, organogenesis, *in vitro* rooting	Prasad (2004)
Cryptolepis buchanani	Shoot	Immature seeds	Multiple shoots, rooted *in vitro*	Prasad (2004)
Cryptostegia grandiflora	Callus	Stem pith cells	Copious proliferation of parenchymatous cells	Synder (1955)
Cynanchum callialatum	Shoot	Root	Direct plantlet regeneration from root explant	Raghuramulu (2001)
Cynanchum callialatum	Callus	Leaf, cotyledon, Internode, root	Callus induction, shoot regeneration, propagation of plantlets	Raghuramulu (2001)
Cynanchum callialatum	Callus	Immature embryos	Somatic embryogenesis, plantlet regeneration	Raghuramulu (2001)
Cynanchum callialatum	Shoot	Node	Multiple shoot induction, rooted *in vitro*	Raghuramulu *et al.* (2002)

contd...

... *contd.* Table 1

Species name	*Type of culture*	*Source of Explant*	*Result*	*Reference*
Cynanchum nigrum (Swallow's wort)		Unpollinated ovaries	Embryogenesis nucellar plantlets	Haccius and Hausner (1976)
Cynanchum vincetoxicum		Unpollinated ovaries	Somatic embryogenesis, embryo germination, plantlets	Haccius and Hausner (1976)
Cynanchum wilfordii	Callus/ Suspension	Stem	Callus, suspension culture, gagaminine synthesis	Shin *et al.* (2003)
Decalepis arayalpathra	Root	Leaf, internode	Root induction	Sudha and Seeni (2001)
Decalepis arayalpathra	Shoot	Node	Multiple shoots, shoots rooted	Gangaprasad *et al.* (2005)
Decalepis arayalpathra	Shoot	Node	Mass propagation, rooted *in vitro*	Sudha *et al.* (2005)
Decalepis hamiltonii	Shoot	Axillary bud	Multiple shoots, *in vitro* rooted	Bais *et al.* (2000a)
Decalepis hamiltonii	Root	Axillary bud	Multiple shoots, effect of $AgNO_3$ on *in vitro* rooting	Bais *et al.* (2000b)
Decalepis hamiltonii	Shoot	Leaf	Shoot bud regeneration, subsequent propagation of the plantlets	George *et al.* (2000)
Decalepis hamiltonii	Shoot	Axillary bud	*In vitro* rooting form regenerated shoots	Obul Reddy *et al.* (2001)
Decalepis hamiltonii	Shoot	Shoot tips	Triacontanol increased shoot multiplication and rooting	Obul Reddy *et al.* (2002)
Decalepis hamiltonii	Shoot	Node	Multiple shoot induction, plantlet propagation	Anitha and Pullaiah (2002)
Decalepis hamiltonii	Shoot	Node	Multiple shoots, rooted *in vitro*	Giridhar *et al.* (2003)
Decalepis hamiltonii	Shoot	Axillary bud	Multiple shoots, *in vitro* rooted, production of 2-hydroxy-4-methoxybenzaldehyde	Giridhar *et al.* (2004)
Decalepis hamiltonii	Callus	Leaf	Somatic embryogenesis, organogenesis, shoot regeneration	Giridhar *et al.* (2004)
Decalepis hamiltonii	Shoot	Node	Multiple shoot induction, shoot elongation, *in vitro* rooting	Gururaj *et al.* (2004)
Decalepis hamiltonii	Shoot	Shoot tip	Shoot multiplication, *in vitro* rooting	Giridhar *et al.* (2005)
Gymnema elegans	Shoot	Nodes either from mature plants or from seedling	Multiple shoot induction, *in vitro* rooting	Komalavalli and Rao (1997)
Gymnema sylvestre	Shoot	Nodal, stem segments	Large scale multiplication of plantlets	Anu *et al.* (1994)

Gymnema sylvestre	Shoot	Mature nodal segment	Formation of multiple shoots, plantlet propagation	Sairam Reddy *et al.* (1998)
Gymnema sylvestre	Shoot	Axillary node, leaf, shoot tip, cotyledonary node, petiole, root, internodal segment	Multiple shoot induction, rooted	Komalavalli and Rao (2000)
Gymnema sylvestre	Callus	Hypocotyl, leaf, cotyledons from seedling	Callus culture, somatic embryos, embryo germination and plantlet formation	Ashok Kumar *et al.* (2002)
Hemidesmus indicus	Callus	Shoot tips, stem, root	Callus proliferation, regeneration of shoot buds, steroid production	Heble and Chadha (1978)
Hemidesmus indicus	Callus	Leaf, stem	Callus proliferation, organogenesis, somatic embryogenesis, plantlet formation	Sarasan *et al.* (1994)
Hemidesmus indicus	Callus	Stem segments	Callus culture, multiple shoot induction, propagation of *in vitro* raised plantlets	Sharma and Yelne (1995)
Hemidesmus indicus	Shoot	Node	Multiple shoot induction, plantlet propagation	Sharma and Yelne (1995)
Hemidesmus indicus	Shoot	Node	Shoot multiplication, rooted *in vitro*	Patnaik and Debata (1996)
Hemidesmus indicus	Root	Root	Production of 2-hydroxy 4-methoxy benzaldehyde	Sreekumar *et al.* (1998)
Hemidesmus indicus	Shoot	Axillary bud	Rapid multiplication, rooted *in vitro*	Malathy and Pai (1998)
Hemidesmus indicus	Shoot	Node	Rapid micropropagation, production of 2-hydroxy 4-methoxy benzaldehyde	Sreekumar *et al.* (2000)
Hemidesmus indicus	Shoot	Node	Multiple shoot induction, plantlet propagation	Raghuramulu (2001)
Hemidesmus indicus	Callus	Immature embryos	Somatic embryogenesis, plantlet regeneration	Raghuramulu (2001)
Hemidesmus indicus	Shoot	Root, cotyledons, internode	Callus proliferation, shoot bud induction, propagation of plantlets	Raghuramulu (2001)
Hemidesmus indicus	Shoot	Node	Shoot bud culture, propagation of plantlets	Saha *et al.* (2003)
Hemidesmus indicus	Shoot	Shoot tip, node	*In vitro* clonal propagation	Neeta *et al.* (2003)
Hemidesmus indicus	Shoot/ callus	Leaves, root	Shoot and callus culture, production of antioxidants	Neeta *et al.* (2005)
Hemidesmus indicus	Callus	Leaf	Induction of callus. somatic embryogenesis and regeneration	Swaroopa and Dixit (2006)

contd...

... *contd*. Table 1

Species name	*Type of culture*	*Source of Explant*	*Result*	*Reference*
Holostemma ada-kodien	Hairy root	Shoot buds, hypocotyl	Genetic transformation, hairy root induction	Karmarkar and Keshavachandran (2001)
Holostemma ada-kodien	Callus	Leaf, node, internode	Callus proliferation, regeneration of plantlets, propagation of plantlets	Martin (2002)
Holostemma ada-kodien	Shoot	Axillary buds	Axillary bud multiplication, plant propagation	Martin (2002)
Holostemma ada-kodien	Callus	Leaf, internode, root	Callus induction, indirect somatic embryogenesis	Martin (2003)
Holostemma annulare	Shoot	Node	Shoot regeneration, plantlet propagation	Sudha *et al.* (1998)
Holostemma annulare		Shoot tips	Cryopreservation of shoot tips	Decruse *et al.* (1999)
Holostemma annulare	Shoot	Root segments	Regeneration of plants from *in vitro* root segments	Sudha *et al.* (2000)
Holostemma annulare	–	Shoot tips	NH_4NO_3 influence on regeneration potential of cryopreserved shoot tips	Decruse and Seeni (2002)
Holostemma annulare	Shoot	Node, shoot tip	Shoot multiplication, shoot Cryopreservation	Decruse *et al.* (2004)
Hoya carnosa var. *compacta*	Callus	Leaf	Callus proliferation, asexual embryogenesis, development of asexual embryos	Maraffa *et al.* (1981)
Hoya carnosa var. *rubra*	Callus	Leaf	Callus proliferation, asexual embryogenesis, development of asexual embryos	Maraffa *et al.* (1981)
Leptadenia reticulata	Callus	Leaf	Indirect somatic embryogenesis, plant regeneration	Hariharan *et al.* (2002)
Leptadenia reticulata	Root	Meristem	Adventitious roots, HPTLC	Bhatt *et al.* (2002)
Leptadenia reticulata	Shoot	Node	Micropropagation	Arya *et al.* (2003)
Leptadenia reticulata	Callus	leaf, internode, node and shoot tip	Indirect somatic embryogenesis, suspension cultures	Martin (2004)
Pergularia pallida	Protoplast	Callus tissue from stem segments	Isolation of protoplast, plant regeneration, propagation	Bapat *et al.* (1986)
Pergularia pallida	Protoplast	Callus	Protoplast isolation, plant regeneration, propagation	Mhatre *et al.* (1987)
Sarcostemma intermedium	Shoot	Node, shoot tip	Axillary bud proliferation, shoot multiplication, rooted *in vitro*	Prasad (2004)
Sarcostemma intermedium	Callus	Internode, root	Callus culture, shoot regeneration	Prasad (2004)

Solanostemma argel	Callus	–	Callus culture, shoot bud proliferation, plant propagation	Amariei *et al.* (1991)
Telosma cordata	Callus	Stem	Embryogenesis, plantlet from germinated embryoids	Prabhudesai and Narayanaswami (1974)
Tylophora asthmatica	Shoot	Node	Multiple shoot induction, plantlet propagation	Usha *et al.* (2002)
Tylophora indica	Callus	Stem segments	Callus culture, somatic embryogenesis	Rao *et al.* (1970)
Tylophora indica	Callus	Stem segments	Embryogenesis, plantlets from geminated embryoids	Rao and Narayanaswamy (1972)
Tylophora indica	Callus	Stem segments	Callus culture, secondary metabolites	Benjamin and Mulchandani (1973)
Tylophora indica	Callus	Stem	Callus culture	Benjamin and Mulchandani (1976)
Tylophora indica	Callus segments	Leaf, stem, root	Callus proliferation, regeneration of shootlets, detection of phenanthroindolizidine alkaloids	Benjamin *et al.* (1979)
Tylophora indica	Protoplast	Callus from stem segments	Protoplast isolation, plantlet regeneration from stem callus	Mhatre *et al.* (1984)
Tylophora indica	Protoplast	Stem segments	Protoplast isolation, plant regeneration	Rao *et al.* (1986)
Tylophora indica	Protoplast	Callus from stem segments	Protoplast isolation, regeneration of shoots from callus	Mhatre *et al.* (1987)
Tylophora indica	Shoot	Nodal stem segments	Induction of axillary buds with ascorbic acid	Sharma and Chandel (1992)
Tylophora indica	Shoot	Axillary bud	NH_4NO_3 influenced shoot number and alkaloid synthesis	Chattopadhyay *et al.* (1992)
Tylophora indica	Shoot	Node	Multiple shoot induction, plantlet propagation	Bera and Roy (1993)
Tylophora indica	Shoot	Leaf	Direct shoot regeneration, plantlet propagation	Bera and Roy (1993)
Tylophora indica	Callus	Leaf	Somatic embryogenesis, estimation of protein, DNA, RNA in embryogenic and non embryogenic cells	Bera and Roy (1995)
Tylophora indica	Shoot	Node	*In vitro* mass multiplication	Sharma *et al.* (1995)
Tylophora indica	Callus	Leaf	Somatic embryogenesis, plant regeneration	Manjula *et al.* (2000)
Tylophora indica	Callus	–	Somatic embryogenesis, RAPD analysis of regenerated plants	Jayanthi and Mandal (2001)
Tylophora indica	Callus	Leaf	Adventitious shoot regeneration	Faisal and Anis (2003)

contd...

... *contd*. Table 1

Species name	*Type of culture*	*Source of Explant*	*Result*	*Reference*
Tylophora indica	Shoot/ Callus	Root	Direct shoot bud formation, somatic embryogenesis, plantlet propagation	Chaudhuri *et al*. (2004)
Tylophora indica	Callus	Stem segments	Callus proliferation, multiple shoot induction, shoots rooted	Faisal and Anis (2005)
Tylophora indica	Callus	Petiole	Callus production, shoot multiplication,	Faisal *et al*. (2005)
Tylophora indica	Callus	Leaf	Organogenic callus formation, shoot regeneration	Thomas and Philip (2005)
Tylophora indica	Transfor-mation/callus	Intact shoots	Genetic transformation, synthesis and secretion of tylophorine	Chaudhuri *et al*. (2005)
Utleria salicifolia	Shoot	–	Micropropagation	Gangaprasad *et al*. (2003)
Wattakaka volubilis	Shoot	Node, shoot tip	*In vitro* propagation	Emmanuel *et al*. (2001)
Wattakaka volubilis	Shoot	Node, shoot tips	Multiple shoot induction, plantlet propagation	Chakradhar (2004)
Wattakaka volubilis	Callus	Root, leaf, internode	Callus proliferation, shoot regeneration, plant propagation	Chakradhar (2004)
Wattakaka volubilis	Shoot	Mature and aseptic seedling node	Shoot multiplication, *in vitro* rooting	Chakradhar and Pullaiah (2006)

References

Ahmed, A.A.B., T. Gowthaman, A.S. Rao and M.V. Rao. 2003. Somatic embryogenesis in *Gymnema sylvestre* R. Br. by suspension culture. *In: National seminar on biodiversity conservation and commercial exploitation*. Abst. No. 87.

Amariei, D., U. Stanescu, E. Gille and T. Onisei. 1991. The biosynthetic capacity of regenerated *Solanostemma argel* (Sel.) Hayne: Callus and shoots. *Revue Roumaine De Biolgie Serie De Biologie Vegetale.*, **36:**71–76.

Ananthan, R. 2003. *In vitro* approaches for the propagation and conservation of the endemic *Ceropegia barnesii* Brance *et* Chattarjee and *Ceropegia pusilla* Wight et Arn. Asclepiadaceae. Ph.D. Thesis, Bharathiyar University, Coimbatore, India.

Anitha, S. 2001. *In vitro* propagation studies of *Decalepis hamiltonii* Wight & Arn. and *Sterculia foetida* Linn. Ph.D. Thesis submitted to Sri Krishnadevaraya University, Anantapur, Andhra Pradesh, India.

Anitha, S. and T. Pullaiah. 2002. *In vitro* propagation of *Decalepis hamiltonii. J. Trop. Med. Plants*, **3:**227–232.

Anu, K.I., P.A. Nazeem, L. Joseph and N. Vijay Kumar. 1994. Response of 'Gurmar' *(Gymnema sylvestre* R. Br.) for *in vitro* propagation. *South Ind. Hort.*, **42:**365–368.

Arya, V., N.S. Shekhawat and R.P. Singh. 2003. Micropropagation of *Leptadenia reticulata* — A medicinal plant. *In vitro Cell Dev. Biol. Plant.*, **39:**180–185.

Ashok Kumar, H.G., Murthy, H.N. and Paek, K.Y. 2002. Somatic embryogenesis and plant regeneration in *Gymnema sylvestre. Plant Cell Tiss. Org. Cult.* **71**: 85–88.

Bais, H.P., J. George and G.A. Ravishankar. 2000a. *In vitro* propagation of *Decalepis hamiltonii* Wight & Arn., an endangered shrub, through axillary bud cultures. *Curr. Sci.*, **79:**408–410.

Bais, H.P., G. Sudha, B. Suresh and G.A. Ravishankar. 2000b. Silver nitrate influence *in vitro* root formation in *Decalepis hamiltonii* Wight & Arn. *Curr. Sci.*, **79:**894–898.

Bapat, V.A., M. Mhatre and P.S. Rao. 1986. Regeneration of plantlets from protoplast cultures of *Pergularia pallida. J. Plant. Physiol.*, **124:**413–417.

Beena, M.R. and K.P. Martin. 2003. *In vitro* propagation of the rare medicinal plant *Ceropegia candelabrum* L. through somatic embryogenesis. *In vitro Cell. Dev. Biol. Plant,* **39:**510–513.

Beena, M.R., Martin, P.B. Kirti and M. Hariharan. 2003. Rapid *in vitro* propagation of medicinally important *Ceropegia candelabrum. Plant Cell Tiss. Org. Cult.*, **72:**285–289.

Benjamin, B.D. and N.B. Mulchandani. 1973. Studies in biosynthesis of secondary constituents in tissue cultures of *Tylophora indica. Planta Med.*, **23:**394–397.

Benjamin, B.D. and N.B. Mulchandani. 1976. Effect of gamma irradiation on biosynthetic potential of callus cultures of *Tylophora indica. Planta Med.*, **29:**37–40.

Benjamin, B.D., M.R. Heble and M.S. Chadha. 1979. Alkaloid synthesis in tissue cultures and regenerated plants of *Tylophora indica* Merr. (Asclepiadaceae). *Z. Pflanzen Physiol.*, **92:**72–84.

Bera, T.K. and S.C. Roy. 1993. Micropropagation of *Tylophora indica* (Burm. f.) Merr. by multiple bud formation from mature leaf explants without interveining callus. *Botanical Bulletin Academia Sinica.*, **34:**83–87.

Bera, T.K. and S.C. Roy. 1995. Nucleic acids and protein synthesis during early stages of somatic embryogenesis in *Tylophora indica. Cell and Chromosome Res.*, **18:**59–66.

Bhatt, T., V. Jain, M.G. Jayathirtha, G. Banerjee and S.H. Mishra. 2002. *In vitro* regeneration of roots of *Phyla nodiflora* and *Leptadenia reticulata*, and comparison of roots from cultured and natural plants for secondary metabolites. *Indian J. Exp. Biol.*, **40:**1382–1386.

Biesboer, D.D. 1983. The detection of cells with a laticifer-like metabolism in *Asclepias syriaca* L. suspension cultures. *Plant Cell Rep.*, **2:**137–139.

Biesboer, D.D. 1984. Changes in protoplast synthesis during somatic embryogenesis in *Asclepias tuberosa* L. tissue cultures. *Am. J. Bot.*, **71:**121–125.

Biesboer, D.D. and J.L. Tate. 1988. Culture of *Asclepias syriaca, A. tuberosa* and *Daucus carota* in ventex cell lift bioreactors. *Abst. Pap. Am. Chem. Soc.*, 195 meet., MBTD 28.

Chakradhar, T. 2004. *In vitro* culture, physiological, phytochemcial and antimicrobial studies of a medicinal plant — *Wattakaka volubilis* (L.f) Stapf. Ph.D. Thesis submitted to S.K. University, Anantapur, Andhra Pradesh, India.

Chakradhar, T. and T. Pullaiah. 2006. Effect of explant source on axillary shoot multiplication during micropropagation of a rare medicinal plant- *Wattakaka volubilis* (L. f) Stapf. *J. Plant Biochem. Biotech.*, **15**:43–45.

Chattopadhyay, S., S.K. Datta and M. Ray. 1992. *In vitro* effect of NH_4NO_3 on growth and alkaloid content of *Tylophora indica* Merr. *Phytomorphology*, **42**:134–144.

Chaudhuri, K.N., B. Ghosh and S. Jha. 2004. The root: a potential; new source of competent cells for high-frequency regeneration in *Tylophora indica. Plant Cell Rep.*, **22**:731–740.

Chaudhuri, K.N., B. Ghosh, D. Tepfer and S. Jha. 2005. Genetic transformation of *Tylophora indica* with *Agrobacterium rhizogenes* A4: growth and tylophorine productivity in different transformed root clones. *Plant Cell Rep.*, **24**:25–35.

Dasgupta, M., T.K. Pramanik and S.K. Datta. 1986. Mass propagation and genetic variability of two cardenolide plants *in vitro. Hort. Sci.* 21: Sect 211. 784.

Datta, S.K. 1981. *In vitro* study of rhizogenesis from shoot tips of *Calotropis gigantea* (Linn.). R. Br. ex Ait. *Indian J. Exp. Biol.*, **19**:566–567.

De Bagga, S., S. Chattopadhyay, A. Kumar, K.L. Ghosh, K. Datta and S.K. Datta. 1986. Differentiation of laticifer in *Calotropis gigantea* hormone and enzymatic regulation of tracheary elements in tissue culture of some plants. *Intl. Congr. Plant Tissue Cell Cult.*, 6 meet. 164.

De, S. 1989. DNA, RNA and protein contents during rhizogenesis in tissue culture of *Calotropis gigantea* (Linn.) R. Br. ex Ait. *Ind. J. Exp. Biol.*, **27**:185–186.

Decruse, S.W. and S. Seeni. 2002. Ammonium nitrate in the culture medium influences regeneration potential of cryopreserved shoot tips of *Holostemma annulare. Cryo Letters*, **23**:55–60.

Decruse, S.W., S. Seeni and G.M. Nair. 2004. Preparative procedures and culture medium affect the success of cryostorage of *Holostemma annulare* shoot tips. *Plant Cell Tiss. Org. Cult.*, **76**:179–182.

Decruse, S.W., S. Seeni and P. Pushpangadan. 1999. Cryopreservation of alginate coated shoot tips of *in vitro* grown *Holostemma annulare* (Roxb.) K. Schum., an endangered medicinal plant: influence of preculture and DMSO treatment on survival and regeneration. *Cryo Letters*, **20**:243–250.

Dhir, S.K. and S.C. Maheshwari. 1986. High yield preparation of viable protoplasts from leaves of *Calotropis procera* (Ait.) R. Br. *Int. Congr. Plant Tiss. Cell Cult.*, **6meet**:49.

Dhir, S.K., N.S. Shekhawat, S.D. Purohit and H.C. Arya. 1984. Development of laticifer cells in callus cultures of *Calotropis procera* (Ait.). R. Br. *Plant Cell Rep.*, **3**:206–209.

Dunbar, K.B., K.J. Wilson, B.H. Petersen and D.D. Biesboer. 1986. Identification of laticifers in embryoids derived from callus and suspension cultures of *Asclepias* species (Asclepiadaceae). *Am. J. Bot.*, **73**:847–851.

Dutta, S.K. and S. De. 1986. Laticifer differentiation of *Calotropis gigantea* R. Br. ex Ait. in cultures. *Ann. Bot.*, **57**:403–406.

Emmanuel, S., K. Kathiravan and S. Ignacimuthu. 2001. *In vitro* propagation of *Wattakaka volubilis* (L.f.) Stapf. *XVI Intl. Bot. Congr.* Abstr., 1653.

Faisal, M. and M. Anis. 2003. Rapid mass propagation of *Tylophora indica* Merrill via leaf callus culture. *Plant Cell Tiss. Org. Cult.*, **75**:125–129.

Faisal, M. and M. Anis. 2005. An efficient *in vitro* method for mass propagation of *Tylophora indica. Biologia Plant*, **49**:257–260.

Faisal, M., S. Singh and M. Anis. 2005. *In vitro* regeneration and plant establishment of *Tylophora indica* (Burm.f.) Merrill: petiole callus culture. *In vitro Cell. Dev. Biol. Plant.*, **41**:511–515.

Gangaprasad, A., S.W. Decruse, S. Seeni and G.M. Nair. 2005. Micropropagation and ecorestoration of *Decalepis arayalpathra* (Joseph & Chandra.) Venter — An endemic and endangered ethnomedical plant of Western Ghats. *Indian J. Biotech.*, **4**:265–270.

Gangaprasad, A., L.G. Nair, K. Radhakrishnan, S. Seeni, G.M. Nair and P. Pushpangadan. 2003. Micropropagation of *Utleria salicifolia* endemic and endangered ethnomedicinal plant of the Western Ghats. *J. Medicinal Aromatic Pl. Sci.*, **25**:19–24.

George, J., H. Bais, G.A. Ravishankar and P. Manilal. 2000. Optimization of media constituents for shoot regeneration from leaf callus cultures of *Decalepis hamiltonii* Wight & Arn. *Hort. Sci.*, **35**:296–299.

Giridhar, P., T. Rajasekharan, S. Nagarajan and G.A. Ravishankar. 2004. Production of 2-hydroxy-4-methoxybenzaldehyde in roots of tissue culture raised and acclimatized plants of *Decalepis hamiltonii* Wight & Arn., an endangered shrub endemic to Southern India and evaluation of its performance *vis-a-vis* plants from natural habitat. *Indian J. Exp. Biol.*, **42**:106–110.

Giridhar, P., D. Vijaya Ramu, B. Obul Reddy, T. Rajasekharan and G.A. Ravishankar. 2003. Influence of phenyl acetic acid on clonal propagation of *Decalepis hamiltonii* Wight & Arn: An endangered shrub. *In vitro cell. Dev. Biol. Plant.*, **39**:463–467.

Giridhar, P., Vinodkumar, D., Obul Reddy, B. and Rajasekharan, T. 2004. Somatic embryogenesis, organogenesis and regeneration from leaf callus culture of *Decalepis hamiltonii* Wight & Arn., an endangered shrub. *In vitro cell. Dev. Biol. Plant.*, **40**:567–571.

Giridhar, P., H, B. Gururaj and G.A. Ravishankar. 2005. *In vitro* shoot multiplication through shoot tip cultures of *Decalepis hamiltonii* Wight & Arn a threatened plant endemic to southern India. *In vitro cell Dev. Biol. Plant.*, **41**:77–80.

Groet, S.S. and G.H. Kidd. 1980. Embryogenesis in callus and suspension cultures of milkweed. *Plant Physiol.*, **65**:Abst. 195.

Gururaj, H.B., P. Giridhar and G.A. Ravishankar. 2004. Efficient clonal propagation for *Decalepis hamiltonii*, an endangered shrub, under the influence of phloroglucinol. *Indian J. Exp. Biol.*, **42**:424–428.

Haccius, B. and G. Hausner. 1976. A transplantable embryogenic callus from nucellar tissue of *Cynanchum vincetoxicum* and the significance of a globular proembryonal cell complex in the non-zygotic embryo development. *Protoplasma*, **90**:265–282.

Hariharan, M., D.P. Sebastian, S. Benjamin and P. Prashy. 2002. Somatic embryogenesis in *Leptadenia reticulata* Wight & Arn. — A medicinal plant. *Phytomorphology*, **52**:155–160.

Heble, M.R. and M.S. Chadha. 1978. Steroids in cultured tissues and mature plant of *Hemidesmus indicus* R. Br. (Asclepiadaceae). *Z. Pflazen Physiol.*, **89**:401–406.

Jayanthi, M. and P.K. Mandal. 2001. Plant regeneration through somatic embryogenesis and RAPD analysis of regenerated plants in *Tylophora indica* (Burm.f.) Merrill. *In vitro Cell Dev. Biol. Plant.*, **37**:576–580.

John Britto, S., E. Natarajan and Arockiasamy. 2003. *In vitro* flowering and shoot multiplication from nodal explants of *Ceropegia bulbosa* Roxb. var. *bulbosa*. *Taiwania*, **48**:106–111.

Karmarkar, S.H. and R. Keshavachandran. 2001. Genetic transformation and hairy root induction in *Holostemma ada-kodien* K. Schum — A vulnerable medicinal plant. *Ind. J. Exp. Biol.*, **39**:1263–1267.

Karpagam, N. and M.V. Rao. 2003. *In vitro* clonal multiplication of *Asclepias curassavica* L.: a multipurpose medicinal herb. *National Symposium on emerging trends in Indian medicinal plants*, 20.

Kaur, G., T.S. Rathore, S. Rama Rao and N.S. Shekhawat. 1992. *In vitro* micropropagation of *Caralluma edulis* (Edgew.) Benth. & Hook.f — a rare edible plant species of Indian desert. *Ind. J. Plant Gen. Res.*, **5**:51–56.

Kedage, V.V., M. Mhatre and G.B. Dixit. 2006. *In vitro* propagation of *Ceropegia noorjahani* Ans.: A critically endangered, endemic medicinal plant of Maharashtra. *International symposium on frontiers in Genetics and Biotechnology — Retrospect and Prospects*, 162.

Knorr, D. and S.M. Miazga. 1987. Production of protease from cell cultures of common milkweed *(Asclepias syriaca* L.). *J. Agric. Food Chem.*, **35**:621–624.

Komalavalli, N. and M.V. Rao. 1997. *In vitro* micropropagation of *Gymnema elegans* Wt. & Arn. — a rare medicinal plant. *Indian J. Exp. Biol.*, **35**:1088–1092.

Komalavalli, N. and M.V. Rao. 2000. *In vitro* micropropagation of *Gymnema sylvestre* — A multipurpose medicinal plant. *Plant Cell Tiss. Org. Cult.*, **61**:97–105.

Lee, C.W. and J.C. Thomas. 1985. Propagation of desert milkweed by shoot tip Culture. *Hort. Sci.*, **20**:263–264.

Lee, C.W., J. Yeckes and J.C. Thomas. 1982. Tissue culture propagation of *Euphorbia lathyris* and *Asclepias erosa*. *Hort. Sci.*, **17**:533.

Malathy, S. and J.S. Pai. 1998. *In vitro* propagation of *Hemidesmus indicus*. *Fitoterapia*, **69**:533–536.

Manjula, S., A. Job and G.M. Nair. 2000. Somatic embryogenesis from leaf derived callus of *Tylophora indica* (Burm.f.) Merrill. *Indian J. Exp. Biol.*, **38**:1069–1072.

Mante, S., H.B. Tepper and R. Scorza. 1989. The dicot cotyledon as a source of adventitious shoots. *In vitro*, **25**:Pt. 2,27A.

Maraffa, S.B., W.R. Sharp, H.K. Tayama and T.A. Fretz. 1981. Apparent asexual embryogenesis in cultured leaf sections of *Hoya*. *Z. Pflanzen Physiol.*, **102**:45–55.

Martin, K.P. 2002. Rapid propagation of *Holostemma ada-kodien* Schult, a rare medicinal plant, through axillary bud multiplication and indirect organogenesis. *Plant Cell Rep.*, **21**:112–117.

Martin, K.P. 2003. Plant regeneration through somatic embryogenesis on *Holostemma ada-kodien*, a rare medicinal plant. *Plant Cell Tiss. Org. Cult.*, **72**:79–82.

Martin, K.P. 2004. Benzyladenine induced somatic embryogenesis and plant regeneration of *Leptadenia reticulata*. *Biologia Plant.*, **48**:285–288.

Mhatre, M., V.A. Bapat and P.S. Rao. 1984. Plant regeneration in protoplast cultures of *Tylophora indica*, *J. Plant Physiol.*, **115**:231–235.

Mhatre, M., V.A. Bapat and P.S. Rao. 1987. Plant regeneration in protoplast cultures of *Tylophora indica* and *Pergularia pallida*. Proc. Symp. Plant Cell and tissue culture of economically important plants. G.M. Reddy (ed.), 49–53.

Neeta, M., M. Pratibha, S.K. Datta, and M. Shanta. 2003. Improvement in clonal propagation of *Hemidesmus indicus* R.Br. through adenine sulphate. *J. Plant. Biotechnol.*, **5**:239–244.

Neeta, M., M. Pratibha, S.K. Datta and M. Shanta. 2005. *In vitro* biosynthesis of antioxidants from *Hemidesmus indicus* R. Br. cultures. *In vitro Cell Dev. Biol. Plant.*, **41**:285–290.

Nikam, T.D. and R.S. Savant. 2006. *In vitro* propagation of the endangered Asclepiad *Ceropegia oculata* Hook. through nodal culture. *International symposium on frontiers in Genetics and Biotechnology - Retrospect and Prospects.*, p. 173.

Obul Reddy, B., P. Giridhar and G.A. Ravishankar. 2001. *In vitro* rooting of *Decalepis hamiltonii* Wight & Arn., an endangered shrub, by auxin and root-promoting agents. *Curr. Sci.*, **81**:1479–1482.

Obul Reddy, B., P. Giridhar and G.A. Ravishankar. 2002. The effect of Triacontanol on micropropagation of *Capsicum frutescens*. L and *Decalepis hamiltonii* W & A. *Plant Cell Tiss. Org. Cult.*, **71**:253–258.

Patil, V.M. 1998. Micropropagation of *Ceropegia* spp. *In vitro Cell Dev. Biol. Plant.*, **34**:240–243.

Patnaik, J. and B.K. Debata. 1996. Micropropagation of *Hemidesmus indicus* (L.) R. Br. through axillary bud culture. *Plant Cell Rep.*, **15**:427–430.

Prabhudesai, V. and S. Narayanaswamy. 1974. Organogenesis in tissue cultures of certain *Asclepiads*. *Z. Pflanzen Physiol.*, **71**:181–185.

Pramanik, T.K. and S.K. Datta. 1986. Plant regeneration and ploidy variation in culture derived plants of *Asclepias curassavica* L. *Plant Cell Rep.*, **5**:219–222.

Pramanik, T.K., S. De, M. Dasgupta and S.K. Datta. 1986. Organ induced cardenolides biosynthesis in cultured tissues of *Asclepias curassavica, Calotropis gigantea* and *Thevetia peruviana*. *Intl. Congr. Plant Tiss. Cell. Cult.*, 6meet 115.

Prasad, P.J.N. 2004. *In vitro* studies of *Cryptolepis buchanani* Roem. & Schult and *Sarcostemma intermedium* Dcne. (Asclepiadaceae). Ph.D. Thesis submitted to S.K. University, Anantapur, Andhra Pradesh, India.

Prasad, P.J.N., T. Chakradhar and T. Pullaiah. 2004. Micropropagation of *Cryptolepis buchanani* Roem. & Schult. *Taiwania.*, **49**:57–65.

Raghuramulu, D. 2001. *In vitro* morphogenetic studies of *Hemidesmus indicus* R. Br. and *Cynanchum callialatum* Buch.-Ham. ex Wight (Asclepiadaceae). Ph.D. Thesis submitted to S.K.University, Anantapur, Andhra Pradesh, India.

Raghuramulu, D. and T. Pullaiah. 1999. *In vitro* propagation of *Ceropegia bulbosa* Roxb. from the seedling explants. *In*: P.B. Kavi kishore (ed.) Plant Tiss. Cult. Biotechnol. Emerging Trends, 202–210.

Raghuramulu, D., T. Pullaiah and K. Sri Ram Murthy. 1999. Occurrence of twin seedlings in *Ceropegia bulbosa* Roxb. *Indian For.*, **125**:835–837.

Raghuramulu, D., K. Sri Rama Murthy and T. Pullaiah. 2002. *In vitro* propagation of *Cynanchum callialatum*. *J. Trop. Med. Plants*. **3**:233–238.

Rao, P.S. and S. Narayanaswamy. 1972. Morphogenetic investigations in callus cultures of *Tylophora indica*. *Physiol. Plant*, **27**:271–276.

Rao, P.S., V.A. Bapat, M. Mhatre and G.K. Patel. 1986. Studies on plant regeneration and transformation in protoplast culture of some flowering plants. *Intl. Congr. Plant Tiss. Cell Cult.* 6 meet 51.

Rao, P.S., S. Narayanaswamy and B.D. Benjamin. 1970. Differentiation *ex ovule* of embryos and plantlets in stem tissue cultures of *Tylophora indica*. *Physiol. Plant.*, **23**:140–144.

Roy, A.J. and D.N. De. 1987. *In vitro* plantlet regeneration of the petrocorp *Calotropis gigantea* R.Br. *In*: R.N. Sharma *et al.* (eds.). Proc. 3rd Bio Energy Society Convention, pp. 123–128.

Roy, A.J. and D.N. De. 1988. Regeneration of plantlets from tissue culture of *Calotropis gigantea* R.Br., a latex producing *Asclepiad*. *In vitro*. **24**:Pt. 2.36A.

Roy, A.J. and D.N. De. 1989. Micropropagation through cell suspension culture of latex yielding plant *Calotropis gigantea* (Linn.) R.Br. *In*: H.L. Sharma and M.R Sharma (eds.). Proc. 6th Bioenergy Convention and Symposium, pp. 86–91.

Roy, A.T. and D.N. De. 1990. Tissue culture and plant regeneration from immature embryo explants of *Calotropis gigantea* (Linn.) R.Br. *Plant Cell Tiss. Org. Cult.*, **20**:229–233.

Roy, A.T. and D.N. De. 1991. Cell suspension culture, somatic embryogenesis and plantlet regeneration in *Calotropis gigantea* (Linn.) R.Br. *In vitro.*, 27:pt.2,149A.

Roy, A.T. and D.N. De. 2000. Cell suspension culture and plant regeneration in latex producing plant, *Calotropis gigantea* (Linn.) R.Br. *Plant Cell Tiss. Org. Cult.*, **63**:15–22

Saha, S., M.J. Mukhopadhyay and S. Mukhopadhyay. 2003. *In vitro* clonal propagation through bud culture of *Hemidesmus indicus* (L.) R.Br. — an important medicinal herb. *J. Plant. Biochem. Biotechnol.*, **12**:61–64.

Sairam Reddy, P., G. Ram Gopal and G. Lakshmi Sita. 1998. *In vitro* multiplication of *Gymnema sylvestre* R. Br. — an important medicinal plant. *Curr. Sci.*, **75**:843–845.

Sarasan, V., E.V. Sonia and G.M. Nair. 1994. Regeneration of Indian sarsaparilla, *Hemidesmus indicus* R. Br., through organogenesis and somatic embryogenesis. *Ind. J. Exp. Biol.*, **32**:284–287.

Sharma, N. and K.P.S. Chandel. 1992. Effect of ascorbic acid on axillary shoot induction in *Tylophora indica* (Burm. f.) Merrill. *Plant Cell Tiss. Org. Cult.*, **29**:109–113.

Sharma, N., K.P.S. Chandel and A. Paul. 1995. *In vitro* conservation of threatened plants of medicinal importance. *Indian J. Plant Gen. Res.*, **8**:107–112.

Sharma, P.C. and M.B. Yelne. 1995. Observation on *in vitro* propagation of sarivas *Hemidesmus indicus* R. Br. *Bull. Medico Ethnobotanical Res.*, **16**:129–132.

Shin, G.H., M.G. Chio and D.W. Lee. 2003. Comparative study of the effects of various culture conditions on cell growth and gagaminine synthesis in suspension culture of *Cynanchum wilfordii* (Maxim.) Hemsley. *Biological and Pharmaceutical Bulletin*, **26**:1321–1325.

Singh, R.J. 1984. The regeneration of milk weed (*Asclepias* spp.) plants from hypocotyl callus, cell suspension cultures and hypocotyl protoplasts. *Z. Pflanzenzucht*, **92**:95–101.

Snyder, F.W. 1955. Growth of excised tissues from the stem of *Cryptostegia grandiflora*. *Bot. Gaz.*, **117**:147–152.

Sree Kumar, S., S. Seeni and P. Pushpangadan. 1998. Production of 2-hydroxy 4-methoxy benzaldehyde using root cultures of *Hemidesmus indicus*. *Biotechnol. Letters.*, **20**:631–635.

Sreekumar, S., S. Seeni and P. Pushpangadan. 2000. Micropropagation of *Hemidesmus indicus* for cultivation and production of 2-hydroxy 4-methoxy benzaldehyde. *Plant Cell Tiss. Org. Cult.*, **62**:211–218.

Sudha, C.G. and S. Seeni. 2001. Establishment and analysis of fast growing normal root culture of *Decalepis arayalpathra*, a rare endemic medicinal plant. *Curr. Sci.*, **81**:371–374.

Sudha, C.G., P.N. Krishnan, P. Pushpangadan and S. Seeni. 2005 *In vitro* propagation of *Decalepis arayalpathra*, a critically ethnomedicinal plant. *In vitro Cell Dev. Biol. Plant.*, **41**:648–654

Sudha, C.G., P.N. Krishnan and P. Pushpangadan. 1998. *In vitro* propagation of *Holostemma annulare* (Roxb.) K. Schum., a rare medicinal plant. *In vitro Cell. Dev. Biol.-Plant.*, **33**:57–63.

Sudha, C.G., P.N. Krishnan, S. Seeni and P. Pushpangadan. 2000. Regeneration of plants from *in vitro* root segments of *Holostemma annulare* (Roxb.) Schum. a rare medicinal plant. *Curr. Sci.*, **78**:503–506.

Suri, S.S. and K.G. Ramawat. 1995. *In vitro* hormonal regulation of laticifer differentiation in *Calotropis procera*. *Ann. Bot.*, **75**:477–480.

Suri, S.S. and K.G. Ramawat. 1996. Effect of *Calotropis* latex on laticifer differentiation in callus culture of *Calotropis procera*. *Biologia Plant.*, **38**:185–190.

Suri, S.S. and K.G. Ramawat. 1997. Extra cellular calcium deprivation stimulates laticifer differentiation in callus cultures of *Calotropis procera*. *Ann. Bot.*, **79**:371–374.

Swaroopa, G. and G.B. Dixit. 2006. Somatic embryogenesis and plant regeneration from leaf cultures of *Hemidesmus indicus* R. Br. a medicinal plant. *International symposium on frontiers in Genetics and Biotechnology — Retrospect and Prospects*, p. 179.

Tepper, H.B. and R.O. Knapp. 1992. Adventitious shoot initiation on excised cultured cotyledons of *Asclepias syriaca*. *Phytomorphology*, **42**:117–123.

Thomas, T.D. and B. Philip. 2005. Thidiazuron-induced high frequency shoot organogenesis from leaf derived callus of a medicinal climber, *Tylophora indica* (Burm. f.) Merrill. *In vitro Cell Dev. Biol. Plant.*, **41**:124–128.

Tideman, J. and J.S. Hawker. 1982. *In vitro* propagation of latex-producing Plants. *Ann. Bot.*, **49**:273–279.

Tideman, J. and J.S. Hawker. 1983. Tissue culture of latex-bearing plants. In: *Proc. Austr. Plant Tissue Culture Conf.*, 2 meet 24.

Torne, J.M., I. Claparols and M.A. Santos. 1992. Somatic embryogenesis in *Araujia sericifera*. *Plant Cell Tiss. Org. Cul.*, **29**:269–274.

Torne, J.M., L. Moysset, I. Claparols and E. Simon. 1996. Photocontrol of somatic embryogenesis and polyamine content in *Araujia sericifera* petals. *Physiol. Plant.*, **98**:413–418.

Torne, J.M., P. Rodriguez, A. Manich, I. Claparols and M.A. Santos. 1997. Embryogenesis induction in petals of *Araujia sericifera*. *Plant Cell Tiss. Org. Cult.*, **51**:95–102.

Usha, M., S. Latha and K. Anil. 2002. Micropropagation of *Tylophora asthmatica* and *Uraria picta*. *Plant Cell Biotechnol. Mol. Biol.*, **3**:73–76.

Vangala, S.K., K.J. Wilson and D.D. Biesboer. 1985. Comparison of protoplast yields from leaves, callus cultures and suspension cultures of *Asclepias* species. *Am. J. Bot.*, **72**:911–912.

Venkateswara, R., K. Sankara Rao and C.S. Vaidyanathan. 1987. Cryptosin — a new cardenolide in tissue culture and intact plants of *Cryptolepis buchanani* Roem. & Schult. *Plant Cell. Rep.*, **6**:291–293.

Vyapari, S., H. Khatamian and M.L. Albrecht. 1993. *In vitro* regeneration in *Asclepias tuberosa* L. *Hort. Sci.*, **28**:508.

Wilson, K.J. and K.B. Dunbar. 1987. Histochemical observations of embryogenesis in suspension cultures of *Asclepias tuberosa* (Asclepiadaceae). *Am. J. Bot.*, **74**:700.

Wilson, K.J. and P.G. Mahlberg. 1977. Investigations of Laticifer differentiation in tissue cultures derived from *Asclepias syriaca* L. *Ann. Bot.*, **41**:1049–1054.

Zafar, R. and Ranju Rana. 2004. Cardiac glycosides in callus culture of *Calotropis procera*. *Indian J. Nat. Prod.*, **20**:27–30.

Index

A

B

C

D

E

G

H

I

L

Venkateswara, R., K. Sankara Rao and C.S. Vaidyanathan. 1987. Cryptosin — a new cardenolide in tissue culture and intact plants of *Cryptolepis buchanani* Roem. & Schult. *Plant Cell. Rep.*, **6**:291–293.

Vyapari, S., H. Khatamian and M.L. Albrecht. 1993. *In vitro* regeneration in *Asclepias tuberosa* L. *Hort. Sci.*, **28**:508.

Wilson, K.J. and K.B. Dunbar. 1987. Histochemical observations of embryogenesis in suspension cultures of *Asclepias tuberosa* (Asclepiadaceae). *Am. J. Bot.*, **74**:700.

Wilson, K.J. and P.G. Mahlberg. 1977. Investigations of Laticifer differentiation in tissue cultures derived from *Asclepias syriaca* L. *Ann. Bot.*, **41**:1049–1054.

Zafar, R. and Ranju Rana. 2004. Cardiac glycosides in callus culture of *Calotropis procera*. *Indian J. Nat. Prod.*, **20**:27–30.

Index